CALIFORNIA SERIES ON SOCIAL CHOICE AND POLITICAL ECONOMY

Edited by Brian Barry, Robert H. Bates, and Samuel L. Popkin

NESTED GAMES

NESTED GAMES

Rational Choice in Comparative Politics

GEORGE TSEBELIS

UNIVERSITY OF CALIFORNIA PRESS
BERKELEY   LOS ANGELES   OXFORD

University of California Press
Berkeley and Los Angeles, California

University of California Press, Ltd.
Oxford, England

© 1990 by
The Regents of the University of California

Library of Congress Cataloging-in-Publication Data

Tsebelis, George.
    Nested games: rational choice in comparative politics/
George Tsebelis.
        p.   cm.—(California series on social choice and political
    economy; 18)
        Bibliography: p.
        Includes index.
        ISBN 0–520–06732–0 (alk. paper)
        1. Comparative government.   2. Social choice.   3. Game
theory.
I. Title.   II. Series.
JF51.T74   1990
320'.01'13—dc20   89-33580
                                                              CIP

Printed in the United States of America
1 2 3 4 5 6 7 8 9

Chapter 6 is reprinted from the *Journal of Theoretical Politics*
with the permission of Sage Publications Ltd.
Parts of chapter 7 are reprinted from the *British Journal of
Political Science* with the permission of Cambridge University
Press.

To my mother, Persefoni

Contents

List of Figures

List of Tables

Acknowledgments

I have been working on this book for several years. Over these years, I have had the privilege to be in different universities: Washington University, Stanford, Duke, and the University of California, Los Angeles. UCLA gave me the necessary time to finish the project, and the Academic Senate and the International Studies and Overseas Program provided me with valuable financial aid, which made my task a lot easier.

Over these years and in all these universities, I had the good fortune to enjoy the help, the challenges, and the converging or diverging comments of numerous colleagues. Without their help, this book would not exist (at least in its present form).

Robert Bates, as series editor, senior colleague, and friend, holds the world record for successful recommendations (that is, recommendations made and accepted) for a manuscript. He tirelessly read *Nested Games* several times, each time making important remarks. As far as I am concerned, it was a very profitable collaboration. I am grateful to him for his advice and support.

I would like to thank James DeNardo, Miriam Golden, Peter Lange, and Michael Wallerstein for reading successive drafts over and over again and giving me their advice (sometimes over and over again) with infinite tolerance.

James Booth, John Freeman, Jeffry Frieden, Geoffrey Garrett, Herbert Kitchelt, and Robert Putnam read what I had several times considered the final draft of the manuscript and persuaded me that one more iteration was necessary.

Numerous people read separate chapters or gave comments when I presented them in various conferences: Arun Agrawal,

Robert Axelrod, Shaun Bowler, David Brady, Pamela Camerra-Rowe, John Ferejohn, Clark Gibson, Bernard Grofman, Virginia Haufler, Brian Humes, Shanto Iyengar, Paul Johnson, William Keech, Edward Lehoucq, Jeannette Money, Thomas Rochon, Susan Rose-Ackerman, Barbara Salert, James Scott, Teddy Seidenfeld, Kenneth Shepsle, David Soskice, John Sprague, and Sven Steinmo.

I would like to acknowledge the help of Mary Lamprech and Sylvia Stein of the University of California Press. Their work greatly improved the quality of the book.

Participation in a conference on connected games organized by Fritz Scharpf in the Max-Planck-Institut fur Gesellschaftsforschung in Cologne, West Germany, helped me clarify my ideas about the theoretical foundations of nested games.

Most of Chapter 6 was published in the *Journal of Theoretical Politics* under the title "Elite Interaction and Constitution Building in Consociational Democracies." It is reprinted here with the permission of Sage Publications Ltd. Parts of Chapter 7 were published in the *British Journal of Political Science* under the title "The Cohesion of French Electoral Coalitions." These parts are reprinted here with the permission of Cambridge University Press. I would like to thank Elinor Ostrom and Ivor Crewe, the editors of these journals, as well as the anonymous referees for their comments.

Musical inspiration has been provided through the years by Wolfgang Amadeus Mozart. His help was always there when needed.

Chapter One

Nested Games and Rationality

This book analyzes cases in which an actor confronted with a series of choices does not pick the alternative that appears to be the best. In the course of the book, the reader will see that British Labour party activists who consider their standing MP too moderate may vote to replace her, although that choice may lead to the loss of a seat for the Labour party; that Belgian elites who are considered in the consociational literature to be accommodating and compromising in character sometimes initiate political conflict; and that French political parties in certain constituencies do not support their coalition partner, leading their own coalition to defeat.

Why are situations in which an actor chooses an alternative that appears to be against her own interests, or not the best she can do under the existing circumstances, intriguing? Why do they demand explanation? Choices that do not appear to be the best an actor can do are puzzling because most observers assume (at least implicitly) that people try to behave in ways that maximize the achievement of their presumed goals, that is, they make optimal choices. The goal of this book is to provide a systematic, empirically accurate, and theoretically coherent account of apparently suboptimal choices. The following examples illustrate the importance and frequence of apparently suboptimal choices in politics.

I. Some Apparently Suboptimal Choices

Urho Kekkonen was first elected president of Finland in 1956. His presidency was so successful that he occupied the office for twenty-five years. It was, according to Duverger (1978, 63), "the longest and most powerful presidency in Finnish history." What is interesting for our purposes is how this presidency became possible. Therefore, I examine the preferences and behavior of the actors involved in the 1956 Finnish presidential election.

According to Finnish law, presidential elections are conducted by a special three hundred–member electoral college. An election may require two rounds if no candidate gains a majority of the votes. The first two ranking candidates then compete in a second round, assuring a majority vote for the winner.

In 1956, three candidates participated in the first round: the agrarian Urho Kekkonen, the Socialist Karl-August Fagerholm, and the incumbent conservative Juo Kusti Paasikivi. The most challenging opponent for Kekkonen, who had the support of the Communist party, was the conservative Paasikivi. One would expect the Communists to support Kekkonen in the first round with all their fifty-six votes. Instead, only fourteen Communists cast their votes for Kekkonen; the majority (forty-two out of fifty-six) voted for the Socialist candidate. Was this a split inside the Communist party? It was not; the Communists disliked Fagerholm intensely.

Why did most of the Communists choose not to support their preferred candidate, Kekkonen, that is, why did they choose suboptimal behavior? In order to understand the logic of the Communist vote, one must consider the full story of the 1956 election. Paasikivi was eliminated in the first round with 84 votes, against 114 for Fagerholm and 102 for Kekkonen. In the second round, when Kekkonen faced Fagerholm, the Communists voted exclusively for the former. Kekkonen was elected with 151 votes; Fagerholm was defeated with 149.

Although the Communists preferred Kekkonen, they voted for Fagerholm in the first round in order to eliminate the more threatening Paasikivi from the race. The Communists misrepresented their preferences in the first round to promote their most preferred outcome in the second round. The Communists understood that the supposed question of the first round—"which one of the three candidates do you prefer?"—was immaterial. First round voting was a path leading to the second round and to a competition between either Kekkonen and Paasikivi or Kekkonen and Fagerholm. Given that Kekkonen could defeat Fagerholm but not Paasikivi in the last round, his supporters took the necessary steps to assure the final victory of Kekkonen: they eliminated from the last round Kekkonen's most dangerous opponent, Paasikivi.

An actor votes strategically or sophisticatedly (as opposed to sincerely) if in one or more rounds of a series of votes, she votes against her preferences in order to assure a more preferred final outcome. According to this definition, the Communists voted strategically in 1956. Had the Communists voted sincerely, Kekkonen would have received 144 votes in the first round, Paasikivi, 84, and Fagerholm, 72. However, in the succeeding round, in which Kekkonen would have faced Paasikivi, Paasikivi would have won the election. Thus, the Communists' behavior, which was surprising at first glance, turns out to be optimal upon closer consideration. It was, in fact, a manifestation of strategic voting.

This is the end of the factual story; however, this is not the end of the conceptual investigation. Farquharson (1969) traced sophisticated voting back to Pliny the Younger, and Gibbard (1973) found that strategic voting is possible in all resolute electoral systems.[1] The possibility of altering outcomes through sophisticated voting leads to a new series of questions. Was strategic voting possible for the Socialists as well as the Communists? If so, could the Socialists have voted in such a way as to prevent Kekkonen from getting elected?

The answer to both questions is affirmative. The Socialists also could have voted strategically and prevented the election of Kekkonen. In fact, if they had withdrawn their candidate in the first or second round, the duel between Kekkonen and Paasikivi would have ended in Kekkonen's defeat, as the Socialists would have wished in such a case. Why didn't they follow such a strategy? If strategic voting for the Communists was not the mistake it appeared at first glance, but rational (that is, optimizing) behavior, and if strategic voting was available to the Socialists, then the Socialists chose a suboptimal option: to vote sincerely. Why?

To vote strategically, Socialist leaders would have had to explain to their own party activists and voters why they were with-

---

1. Resolute electoral systems are those that exclude ties. For a similar proof that does not require resoluteness, see Schwartz (1982).

drawing their quite successful candidate—a difficult task. This constraint meant the Socialist leadership was involved in two different games simultaneously. In the parliamentary arena, where the president of Finland was to be decided, strategic voting was the optimal choice. In the internal (party) arena, however, where maintaining the allegiance of activists and voters was at stake, sophisticated voting was not possible. When the consequences of strategic voting in both arenas were considered together, strategic voting ceased to be optimal.

The situation was different for the Communists for two reasons. First, Kekkonen was not the Communist candidate, but an agrarian one, so the Communists did not have to explain why they did not vote for their own candidate. Second, Communist parties all over the world (at least in 1956) were known for their observance of the principle of "democratic centralism," which prescribes obedience once a decision is made. Democratic centralism minimizes internal discord and provides the leadership with necessary freedom of movement. So, although the Communists also were involved in games in multiple arenas, the constraints in the internal arena were not important, and the optimizing choice in the parliamentary arena was the optimal strategy overall.[2]

This story presents a series of puzzles. In the beginning, the Communists appeared to behave in a suboptimal way. Once their behavior was explained as strategic voting, the question changed to why the Socialists voted sincerely and thus behaved suboptimally. Once the Socialist behavior became intelligible, that is, when it was explained as optimal behavior, then the question shifted to why the two parties behaved differently, why optimal behavior for one party was suboptimal for the other.

The puzzles presented in the Finnish situation are not isolated. Generally, situations of political representation generate simultaneous involvement in several games: in the parliamentary game and in the electoral game for MPs, in a bargaining game and in a leader-follower game for trade union representatives, in an international game and in a domestic politics game for national leaders.

2. At this point, one could ask why the two parties are organized differently and try to explain their organization as an optimal response to different goals or an optimal adaptation to different conditions. But doing so is beyond the scope of this book.

The interaction between economics and politics can also be conceptualized as several different games played by the same actors.

The study of any one of these games in isolation may lead to puzzles similar to the Finnish case. Only the study of the whole network of games in which an actor is involved will reveal the actor's motivations and explain his behavior.

Sometimes an actor's involvement in several games is accidental. Two usually independent games get connected: imagine wage negotiations in some Western country in the 1960s and then in the 1970s. In the first case, the game can be studied in isolation. In the second, the consequences of the 1973 oil shock have to be introduced. At other times, institutions are explicitly designed to alter the results of isolated games. Compare the deliberations of a parliament with the deliberations of a jury or the Supreme Court. In the first case, debates are public, and elections follow at regular intervals. In the second, every measure is taken to isolate the game. In the first case, the input of the public and different pressure groups is structurally assured. In the second, every measure is taken to assure the independence of the players from any consideration outside the game itself. Finally, sometimes the connection between different arenas may itself be part of a political struggle: conservative economists argue for the separation of economic from political games because they believe that free markets produce efficient economic outcomes and government intervention is an impediment to efficiency. Others believe that government action (which may be suboptimal from the purely economic point of view) corrects politically unacceptable outcomes generated by the market. In a general way, one can argue that democracies have built-in situations where games are not played in isolation and, therefore, where choices may appear to be suboptimal.

II. Nested Games: The Logic of Apparently Suboptimal Choice

The assumption that people maximize their goal achievement is not the only possible starting point for an explanation of suboptimal choice. One could argue that Finnish parties make mistakes; that the English activists, the Belgian elites, and the French parties considered in later chapters make mistakes; or that all these political actors were motivated by other forces, such as habit or jealousy; or that Communists or labor activists belong to a different culture. One could also disregard individual actions and argue that such issues are not important, that what matters in political science are general, "systemic" characteristics and not the properties of individuals.

This book does not follow any of these directions. Along with the mainstream of contemporary political science, I assume that human activity is goal oriented and instrumental and that individual and institutional actors try to maximize their goal achievement. I call this fundamental assumption the *rationality assumption*.

Unlike others in the mainstream, however, I make the rationality assumption explicit, derive its consequences, and draw upon it when formulating explanations. Moreover, I assume that at every step of the way, political actors respect the requirements of rational behavior. In this sense, rational action is one of the explicit major themes of this book; in other words, this book is a rational-choice approach to comparative politics.

Chapter 2 enumerates the requirements of rationality. I show that one of these requirements is conformity to the prescriptions of game theory whenever individuals interact with one another. Therefore, I use game theory to study the interactions among different political actors.

Chapter 3 explains the fundamental game theoretic material used in the book. In game theory, the players face a series of options (strategies); when each selects one strategy, the players jointly determine the outcome of the game, receiving the payoffs associated with that outcome. In order to find the solution to a problem, game theory assumes that the rules of the game (which determine the available strategies) and the payoffs of the players are fixed. Once the rules and payoffs are fixed, the actors choose mutually optimal strategies; each player selects a strategy that maximizes his payoff, given what the other players do. This account specifies that game theory does not leave room for suboptimal action.

How can suboptimal action exist? How can an actor with a series of options $A_1, \ldots, A_n$, out of which A_i appears to be optimal, choose something different from A_i?

Cases of apparently suboptimal choice are in fact cases of disagreement between the actor and the observer. Why would the actor and the observer disagree as to what the optimal course of action is? There are two possibilities: either the actor actually does choose a nonoptimal strategy, or the observer is mistaken.

There are two cases in which the actor does choose suboptimally: if he cannot choose rationally,[3] or if he makes a mistake. For reasons I explain in Chapter 2, I do not think the first case is important in the study of political phenomena. The second case cannot occur often because if the actor recognizes that he was mistaken, he will presumably correct his behavior.

There are also two cases in which the observer may not recognize the optimal course of action. First, the observer makes a mistake, thinking that the optimal action is A_i when it is not. Second, the observer thinks the available set of actions is limited to $A_1, \ldots, A_n$ when it is not—some additional options may be available, including one that is better than A_i.

This book studies *apparently* suboptimal actions because they are frequently cases of disagreement between actor and observer. Therefore, I focus on the reasons the observer failed to recognize the optimal action. To summarize, the argument of this book is that if, with adequate information, an actor's choices appear to be suboptimal, it is because the observer's perspective is incomplete. The observer focuses attention on only one game, but the actor is involved in a whole network of games—what I call *nested games*. What appears suboptimal from the perspective of only one game is in fact optimal when the whole network of games is considered.

There are two major reasons for disagreement between actor and observer. First, option A_i is not optimal because the actor is involved in games in several different arenas, but the observer focuses on only one arena. Let us call the arena that attracts the observer's attention the *principal arena*. The observer disagrees with the actor's choices because the former sees the implications of the latter's choices only for the principal arena. However, when the implications in other arenas are considered, the actor's choice is optimal. I refer to this case of nested games as *games in multiple arenas*.

In the second case, option A_i is not optimal because the actor

---

3. I explain the requirements of rational choice in Chapter 2.

"innovates," that is, takes steps to increase the number of available options so that some new option is now better than A_i. Increasing the available options means actually changing the rules of the game that define the options available to each player. In this case, the observer does not see that the actor is involved not only in a game in the principal arena, but also in a game about the *rules* of the game. I call this case of nested games *institutional design*.[4]

Both kinds of nested games (games in multiple arenas and institutional design) may lead to apparently suboptimal choices. In the case of games in multiple arenas, the observer considers the game in the principal arena without taking contextual factors into account, whereas the actor perceives that the game is nested inside a bigger game that defines how contextual factors (the other arenas) influence his payoffs and those of the other players. In the case of institutional design, the game in the principal arena is nested inside a bigger game where the rules of the game themselves are variable; in this game, the set of available options is considerably larger than in the original one. The actor is now able to choose from the new set one strategy that is even better than his best option in the initial set.

An element of surprise is present in all cases of disagreement between actor and observer. The factor that may vary is the intensity or magnitude of the surprise. Sometimes the actor and the observer disagree on details, so the actor appears to make a very small mistake; sometimes the observer thinks a priori that exactly the opposite course of action was appropriate, so the actor appears to make a choice totally against his own interests. From a theoretical point of view, all cases of suboptimal choice are puzzling; from an empirical point of view, only serious disagreements between observer and actor indicate some fundamental misperception by the observer or some important inadequacy of existing theories.

For each of the two kinds of nested games (games in multiple arenas and institutional design), the book makes two essential contributions: one substantive, and one methodological. In the case of games in multiple arenas, any of the actor's moves has

---

4. The reason I use the term *institutional design* instead of *institutional game* will become clear in Chapter 4.

consequences in all arenas; an optimal alternative in one arena (or game) will not necessarily be optimal with respect to the entire network of arenas in which the actor is involved. Although the observer of only *one* game considers some behavior irrational or mistaken, the behavior is in fact optimizing inside a more complicated situation. The actor may choose a suboptimal strategy in one game if this strategy happens to maximize his payoffs when all arenas are taken into account. The substantive contribution of this examination of games in multiple arenas is that it presents a systematic way to take into account contextual factors (the situation in other arenas). Such contextual factors influence the payoffs of the actors in one arena, leading to the choice of different strategies; therefore, the outcomes of the game are different when contextual factors are taken into account.

In the case of institutional design, a rational actor seeks to increase the number of alternatives, thereby enlarging his strategy space. Instead of confining himself to a choice among available strategies, he redefines the rules of the entire game, choosing among a wider set of options. Therefore, institutional changes can be explained as conscious planning by the actors involved. In the case of institutional design, disagreement between the actor and the observer stemmed from the fact that the observer did not anticipate the actor's political innovation. Had the observer known that additional options existed, he would have agreed that one of the new options was optimal. So institutional design provides a systematic way to think about political institutions. Institutions are not considered simply inherited constraints, but possible objects of human activity.

The conventional game theoretic way to deal with problems of games in multiple arenas or institutional design is to consider all the actors involved in all existing arenas, write down all their available strategies, add all the possible innovating strategies, and solve this giant game. In this giant game, all contextual (other relevant actors and arenas) and institutional (rules of possible games) factors are taken into account. If such an enterprise were possible, and if both the actor and the observer were solving this giant game, there would be no possible disagreement about what constitutes optimal action. However, such a heroic enterprise is impossible—at least for practical purposes.

In order to reduce this problem to manageable dimensions and show the reasons for disagreement between actors and observers, I deal with each case of apparently nonoptimal choice (games in multiple arenas and institutional design) separately. I utilize a technically simple model to represent games in multiple arenas. In Chapter 3, I explain the relation between my model and traditional game theoretic approaches. This representation leads to empirically interesting results while keeping the level of mathematical expertise demanded of the reader to high school algebra.

Technically, games in multiple arenas are games with variable payoffs; the game is played in the principal arena, and the variations of the payoffs in this arena are determined by events in one or more other arenas. The nature of the final game changes, depending on the order of magnitude of these payoffs, whether or not the actors can communicate with one another, and whether or not the game is repeated over time.

Technically, institutional change is presented as a problem of intertemporal maximization, where complications arise because future events cannot be clearly anticipated. The available information about future events is of crucial importance for the choice of different types of institutions.

To recapitulate, in the presence of adequate information, if actors do not choose what appears to be the optimizing strategy, it is because they are involved in nested games: games in multiple arenas or institutional design. *Games in multiple arenas* are technically represented by games with variable payoffs. Payoff variations are determined by and reflect contextual factors. The payoffs of the game in the principal arena vary according to the situation prevailing in other arenas, and the actors maximize by taking into account these variable payoffs. The term *institutional design* refers to political innovation concerning the rules of the game. The actors choose among the different possible games, that is, among the possible sets of rules. In this case, they enlarge their strategy space and choose a previously unavailable option.

I indicated that disagreement between actor and observer stems from either a wrong choice by the actor or the incomplete perspective of the observer. If we assume actor rationality, the first case (the less important) is eliminated. The remaining case can be explained by the nested games framework in which choices appear to be suboptimal in one game because the observer does not take into account that the game in the principal arena is nested inside either a network of other arenas or a higher order game where the rules themselves are variable. Within this rational-choice approach, and assuming adequate information, the concept of nested games is the only explanation for the choice of apparently suboptimal strategies.

III. Outline of the Book

The book describes situations in which actors do *not* choose the apparently optimal alternative because they are involved in nested games, that is, contextual or institutional factors have an overriding importance.

The two kinds of nested games (games in multiple arenas and institutional design) in principle require equivalent treatment. In practice, however, there is an asymmetry. I provide a complete theoretical treatment of games in multiple arenas, draw implications from this treatment, and test these implications in different empirical situations. I treat institutional design in a less rigorous way—I draw up a typology of institutions and observe different kinds of institutions in the empirical chapters that fit this typology. I treat institutional design less exhaustively than games in multiple arenas because institutional change by definition involves political innovation, and it is difficult (if not impossible) to know its rules, much less to have a complete theory about them. Riker (1986) considers the development of political innovation an art as opposed to a science, gives it the name *heresthetics*, and argues that its laws are unknowable. Whether the laws of institutional design are unknowable or simply unknown, the issue of institutional design is too important to be left out of a book adopting a rational-choice methodology. However, the current state of knowledge on institutions justifies the absence of theoretic rigor.

This asymmetry of treatment is clear in the difference in theoretical precision between Chapters 3 and 4. Also, for each of the empirical chapters (5, 6, and 7), the effects of context occupy the major part of the exposition, and only the final section discusses the politics of institutional change. Although theoretically each reason for nonoptimal choice deserves equal treatment, in practice, there are a major and a minor theme to the book: in the major theme, institutions are assumed constant, and I focus on the effects of political context (games in multiple arenas). In the minor theme, I study the change of rules (institutional design).

The presentation is organized in the following way: Chapter 2 examines the implications of the rational-choice approach in detail. I show why and how this approach differs from other research programs in the social sciences. The approach entails a series of requirements for political actors: the absence of contradictory beliefs, the absence of intransitive preferences, and conformity to the axioms of probability calculus and the rules of game theory (to name but a few). How realistic is such an approach? Once the range of applicability of the theory is defined, the rational-choice approach is a legitimate and fruitful approximation of reality.

In Chapter 3, I lay out the theoretical foundation of games in multiple arenas: they are games with variable payoffs, where the payoffs of the game in the principal arena are influenced by the situation prevailing in another arena. The chapter examines simple two-by-two games with variable payoffs, providing the basis for subsequent applications. The relationship among familiar games (prisoners' dilemma, chicken, assurance game, and deadlock) is examined and their equilibria identified, familiarizing the reader with their game theoretic properties. The distinction between one-shot and iterated games is introduced, and the differences in outcomes are derived theoretically. Finally, I examine comparative statics results (for example, what happens to the frequency of choice of different strategies when these games are iterated and the payoffs of the players vary). Each empirical chapter presents a different substantive application of the concept of games in multiple arenas in different Western European countries.

Chapter 3 provides the direct theoretical foundation for the subsequent empirical chapters, and I refer frequently to its results. Nontechnical readers could take the references to Chapter 3 on faith. In this case, they may see in this book little more than three empirical chapters with loose connections to one another. It would be much more profitable if they tried to work their way through the elementary mathematics of Chapter 3 to understand the logic of the subsequent arguments. In this case, the unity of the empirical chapters as demonstrations of the logic of nested games will become apparent, and other cases amenable to similar theoretical treatment will become clearer. What is required for complete understanding of the book is not prior knowledge of mathematics, but the will to study Chapter 3 so that its arguments are familiar each time they are used.

Chapter 4 deals with institutional design. It is a study of the necessary conditions for institutional design, a classification of different kinds of institutional design, and a discussion of the conditions under which they are likely to occur. Institutions are divided into efficient (those that promote the interests of all or almost all the actors) and redistributive (those that promote the interests of one coalition against another). The latter is subdivided into consolidating institutions (institutions designed to promote the winners' interests) and new deal institutions (institutions designed to split existing coalitions and transform losers into winners). I argue that theorizing about institutions has usually been confined to only one of these three cases, and has not been extended to all three. Failure to understand the complex nature of institutions has generated incorrect extrapolations and inferences about them. Some authors (mainly Marxists) see institutions exclusively as redistributive; others (mainly economists) see them as exclusively efficient. Finally, I specify the conditions under which efficient or redistributive institution building prevails. Each of the subsequent empirical chapters of the book presents more systematically one example of each category of institution.

I then apply the theoretical framework defined in Chapters 2, 3, and 4 to three different political phenomena in three different countries: party politics and relations between leaders and activists in the British Labour party, consociationalism and institutional design in Belgium, and electoral politics and coalition cohesion in the French Fifth Republic. The cases were selected for their diversity in order to demonstrate the logical coherence, substantive versatility, and empirical accuracy of the nested games framework.

The book as a whole adopts a "most different systems design" (Przeworski and Teune 1970). Three very different cases in Western European politics are studied. They involve different actors, concern different countries, and focus on different subject matters. In all these cases, some simple propositions about rational behavior apply: changes in payoffs or institutions lead actors to modify their choice of (equilibrium) strategies. Consequently, political context and political institutions matter in predictable ways.

The chapters are presented in order of increasing complexity. Chapter 5 focuses on the interaction between masses and elites in a competitive electoral context. The principal game is the interaction between Labour members of Parliament and their constituency activists, and this game is nested inside a game of electoral competition between parties. Chapter 6 adopts the reverse perspective. The principal game is the interaction among elites; this interaction, however, is influenced by the interaction between each political elite and the masses it represents. The principal game is parliamentary, and it is nested inside a game between elites and masses. Chapter 7 deals with the more complicated situation in which four parties are organized in two coalitions, and each party has to take several arenas into account: the game at the national level, the competitive game among coalitions at the constituency level, and the game between partners at the constituency level. With respect to institutional design, Chapter 5 presents the case of redistributive institutions of the new deal type, Chapter 6 demonstrates how efficient institutions work, and Chapter 7 shows how different winning coalitions adopt different consolidating institutions.

Chapter 5 deals with party politics and the relationship between leadership and party activists. Labour party constituencies occasionally revolt against their MPs and replace them for being too moderate. Sometimes, in the subsequent election, Labour loses the seat. Such suicidal behavior is problematic inside a rational-choice framework. The phenomena of readoption conflicts and their destructive consequences are studied as a repeated game between constituency activists, standing MPs, and Labour party leaders, which is nested inside the competitive game between the Conservative and Labour parties at the constituency and national levels. The activists' apparently suicidal behavior is explained as optimal in this nested game because they are concerned with building a reputation for toughness that will deter their representatives from being moderate.

The nested games framework explains why previous empirical studies (particularly studies that try to assess the relative strengths of constituencies and leaderships by examining the frequency of readoption conflicts or their outcomes [Janosik 1968; McKenzie 1964; Ranney 1965, 1968]) focus on the wrong explanatory variables and thus come to dubious conclusions. Moreover, the nested games framework reveals the importance of the institutional changes made under pressure from constituency activists between 1979 and 1981. Contrary to the existing literature (Kogan and Kogan 1982; Williams 1983), I argue that the major change in the Labour party was the shift to the left in the political preferences of the trade unions in the 1970s and not the subsequent institutional modifications that reflected and crystallized this shift.

Chapter 6 deals with the question of consociationalism and institutional design. According to the consociational literature (Lehmbruch 1974; Lijphart 1969, 1977; McRae 1974), deep political and social cleavages do not lead to explosive and unstable situations as long as political elites are accommodating. Other authors (Billiet 1984; Dierickx 1978) claim that what explains the accommodating behavior of elites in consociational countries is the possibility of package deals across issues: for issues of asymmetric importance, vote trading is possible. If these explanations were correct, there would be two consequences. First, there would be no reason for elites to initiate political conflict. Second, there would be no need for consociational institutions, that is, institutions specially designed to minimize conflict. Both the initiation of conflict and consociational institution building seem to be suboptimal activities according to these theories.

In order to explain these puzzles of suboptimal behavior, I use the nested games framework. I study Belgian political elites as they are involved in nested games. They play the parliamentary game among themselves while each elite is involved in a game with its followers. This game between each elite and the masses it represents influences the payoffs of the parliamentary game. I argue that the behavior of political elites is optimal within the nested games, even though it may not be optimal in either game considered in isolation, and I show that optimal behavior in the nested game sometimes entails the initiation of conflict by elites. I provide a consistent explanation of the design of Belgian institutions. Finally, I use the nested games framework to account for the actors' calculations and the failure of the negotiations concerned with the

Egmont Pact, which was intended to resolve the status of Brussels in 1977.

Chapter 7 deals with electoral politics and coalition cohesion in the French Fifth Republic. The French electoral system requires cooperation and coalition formation among different parties in the second round of the elections. Inside each coalition, the party that arrives second in the first round has to transfer its votes to the winner in the second round. How effectively are parties going to transfer their vote to their partner in the second round?

Spatial models of voting and party competition (Bartolini 1984; Rosenthal and Sen 1973, 1977) predict the following: Communists will vote socialist in the second round because Socialists are more to the left than the right-wing parties. But Socialists will not be stable allies for the Communists because the Socialists do not necessarily feel closer to the Communists than to the right-wing parties. Therefore, Socialists enjoy a "positional advantage" over Communists in electoral politics and coalition building (Bartolini 1984, 110). Similar arguments can be made for the right-wing parties. Because their ideological distance is smaller than that between Socialists and Communists, the transfer of votes will be expected to be better inside the Right than inside the Left. However, in reality, all parties intermittently transfer votes. Why would parties prefer to give a seat to the rival coalition instead of helping their partner win?

To explain this suboptimal behavior, I consider the game between partners at the national level as nested inside the competitive game between coalitions and the game between coalition partners at the constituency level. The conditions prevailing at the local level determine each player's payoffs, and the payoffs determine the likelihood of cooperation. The conclusion of the nested games approach is that vote transfers are determined by the balance of forces in a constituency. This balance includes the relative strength of the coalitions and the relative strength of the partners inside each coalition. The theoretical advantage of the nested games approach is that it demonstrates that all parties obey the same laws and behave in similar ways with respect to coalition cohesion and vote transfers. Comparison of the nested games approach with alternative explanations such as spatial models, survey research (Jaffré 1980), and psychosociological approaches

(Converse and Pierce 1986; Rochon and Pierce 1985) indicates several advantages of the approach: theoretical parsimony, consistency with other existing theories, and descriptive accuracy.

The performance of the nested games approach in each case study should not distract readers from the major point: all the empirical cases, which range from coalition politics to party politics and from questions of ideology to questions of institution building, are applications of the *same* theory. The essential goal of this book is to demonstrate that political context and political institutions matter in predictable ways, to explain why such regularities occur, and to provide a *systematic* way to deal with complicated political phenomena. The emphasis is on the word *systematic* because I hope the book makes this particular method of study widely accessible. Making the production of knowledge accessible is, I believe, an important goal for any scientific enterprise.

Chapter Two

In Defense of the Rational-Choice Approach

Rationality, as defined in Chapter 1, is nothing more than an optimal correspondence between ends and means. Because it is difficult to imagine political processes without the means/ends relationship, this definition may seem tautological, innocuous, and trivial to the point that its discussion is unnecessary.

These impressions are false. First, it is not true that the rational-choice approach is the only possible approach to politics. Section I of this chapter reminds readers that the list of alternative approaches is quite long. In particular, theories such as systems theory and structural functionalism are not concerned with actors, and others such as psychoanalysis, social psychology, and behaviorialism do not necessarily consider actors to be rational. Second, my definition of rationality is not innocuous: Section II of this chapter demonstrates that this simple definition of rationality imposes many requirements on actors. In particular, rational actors must be consistent (have no contradictory beliefs or desires), decide according to the rules of probability calculus, and interact with other actors according to the prescriptions of game theory. Consequently, the reasonable question becomes not whether people ever deviate from rationality, but whether people ever conform to it. In fact, most of the objections to the rational-choice approach suggest that the rationality assumption is not trivial, but rather unrealistically demanding; according to these objections, rational actors do not (and probably cannot) exist. Section III addresses these objections. I indicate that there are good reasons why political actors *should* be rational (a normative approach) and additional reasons why political actors can be studied using the rational-choice approach (a positive approach). Section IV enumerates the principal advantages of the rational-choice approach.

I. What the Rational-Choice Approach Is
Not

One can distinguish two broad categories of theories that do not assume any correspondence between means and ends. The first is primarily unconcerned with actors as units of analysis; the second studies actors but does not assume they are rational.

(1) *Theories without actors.* Systems analysis (Easton 1957), structuralism (Holt 1967), functionalism of the Right (Parsons 1951) or of the Left (Holloway and Picciotto 1978), and modernization theories (Apter 1965) are prominent representatives of this approach. Explanations of social or political phenomena are given in holistic terms, in reference to the system as a whole. Although the existence of rational actors is not denied, the study of their decision-making processes is considered secondary or unimportant. Valid explanations are either causal or functional. In other words, processes or structures can be explained either by antecedent processes and structures or by their beneficial consequences for subsequent processes, structures, and the system itself.

Such theories have a different focus of attention from the rational-choice approach. However, a translation from one research program to the other is sometimes possible. For example, economic modernization has political consequences (Kautsky 1971) because it generates economic interests expressed by political coalitions. These coalitions may or may not achieve their goals because of constraints embedded in existing structures or because of the actions of other coalitions. Or the need for political order in Third World countries (Huntington 1968) can be attributed to a specific group of actors (usually elites) and their interest in certain forms of political organization.

These examples indicate that a translation exists between the individual and aggregate levels. A closer examination of the actor's decision-making process may indicate why situations with similar antecedent conditions evolve differently and demonstrates further the fruitfulness of this translation.

There are instances, however, in which these specific translations between research programs are not possible. Consider Coser's (1971) argument: "Conflict within and between bureaucratic structures provides the means for avoiding the ossification and ritualism which threatens their form of organization."[1] There are two possible meanings: the first interpretation is a comparative statics statement in which systems with conflicting bureaucratic structures demonstrate lower degrees of ossification and ritualism than systems with nonconflicting bureaucratic structures; the second interpretation attempts to *explain* the existence of conflict by its function. The former argument can be empirically tested and found true or false. No claim of explanation is made in this interpretation. The phrase "provides the means for avoiding" should be replaced by "has the effect of reducing," and a rational-choice explanation can be sought for this empirical regularity. The second interpretation has no possible translation in rational-choice terms because there is no actor with the implied goal of avoiding ossification and ritualism; "the system" is an abstraction for a set of individuals with different or conflicting interests and goals. As a result, the emergence of conflict cannot be explained in terms of its beneficial consequences for bureaucratic structures; it has to be explained as an aggregation of behaviors undertaken to promote particularistic goals.

I refer to a shortcut or black box explanation whenever a translation can be made from nonactor theories to the rational-choice approach. In this case, in order to emphasize the macropicture, the entire mechanism of a social or political phenomenon will not be completely described. Where such a translation is impossible, no micromechanisms compatible with observed aggregate results are possible, meaning no causal process can account for the phenomenon. Therefore, we are faced with a case of what is known as *spurious correlation*.

The reason this translation across research programs is important is the principle of *methodological individualism*, which

1. See Elster (1983, 59).

asserts that all social phenomena can and should be explained in terms of the actions of individuals operating under prevailing constraints. Elster (1983) claims that this principle is a special case of the reductionism that exists in any science.

Residing between rational actor and nonactor theories are those that derive political outcomes from the actions of informal social aggregates: classes or groups. These aggregates are considered rational (in the means/ends sense I defined in the beginning of the chapter), but their very existence remains unexplained in terms of rationality. Consider economic and social conflict. One could focus on the conflict among different groups of workers or on the conflict among different branches of industry (workers and capitalists taken together). Instead, Marx considered workers and capital as unified actors trying to maximize their respective welfare (aggregate wages for workers and aggregate profits for capitalists). In his approach, class struggle, the motor force of history, results from the fact that output is fixed at any given time and must be divided between capitalists and workers. Note that both labor and capital are considered unitary actors in this approach and that competition among capitalists for markets or competition among workers for jobs is assumed out of the model in its most simplified form.[2] Conflict between branches of industry is absent as well. These problems may be addressed in subsequent developments (Przeworski and Wallerstein 1982, 1988). The results, however, may be radically different from the original theory.

(2) *Theories with nonrational actors.* The source of nonrationality cannot be the goals of the actor—*De gustibus non est disputandum.* Goals may be egoistic or altruistic, idealistic or materialistic. The only source of nonrationality must be a breach in the means/ends relationship of our definition of rationality.

This breach can occur in two different ways: either through impulsive action or through a deeper source of irrationality (Boudon 1986, 294). Inquiry into both kinds of nonrationality originates from psychology and can be placed in two distinctive classes. The

2. In other parts of his work, Marx deals with the problem of multiple players (capitalists and workers) without, however, entering into the interactions among them. The most famous example is the falling rate of profit discussed in *Das Kapital*, which can be modelled as a prisoners' dilemma game between capitalists. See Boudon (1977).

first class includes theories explaining actions resulting from affective or impulsive motives (e.g., revolutions explained through the "relative deprivation" theory) (Gurr 1971). In this class of theories, departures from rational calculations can be observed and explained by the outside observer and accepted by the actor himself. However, such behavior cannot be systematic or even frequent, as I demonstrate below.

In the second class, the motive for nonrational behavior is a theoretical construct, which may be inaccessible to both the observer and the actor. Such theories include the "imitation instinct" (Gabriel Tarde), "false consciousness" (Friedrich Engels), "inconscient pulsions" (Sigmund Freud), "habitus" (Pierre Bourdieu), "national culture" (Gabriel Almond and Sidney Verba), or forces such as "resistance to change" or "inertia."[3]

Again, it may be possible to translate these theories into a rational-choice approach. As this book argues, certain actions may seem nonrational because the frame of reference is not appropriate. For example, Samuel Popkin and Robert Bates, instead of using the concept of "moral economy," as James Scott does, explain customs and behaviors in rural societies through rational-choice arguments.[4] Bhaduri (1976) explains "resistance to change," that is, why peasants in West Bengal resisted technological innovations that would have improved productivity; he claims that such improvements would reduce debt, thus ending the dependence of the poor on the rich. Consequently, rich landlords objected to innovation in order to preserve their long-term interests.

In other cases, rational-choice approaches translate the independent variables of existing studies into dependent variables and explain the findings of other scholars. For example, Boudon produces a simple rational-choice model to explain one of the most puzzling findings of *The American Soldier*: that pilots who belonged to groups that received frequent promotions were unsatisfied, whereas military policemen were satisfied by a system in which promotions were rare.[5] According to Boudon's account, in-

---

3. For a critical examination of some of these theories, see Barry (1978).

4. See Popkin (1979), Bates (1983), and Scott (1976). A different interpretation of Scott's work would be that it provides the structural reasons for the risk averse behavior of peasants. The issue of risk aversion is discussed in the appendix to this chapter.

5. See Boudon (1979). The original findings appeared in Stouffer (1965).

dividuals understand the characteristics of the system and invest their efforts accordingly; if the probability of reward is small, the expected utility from trying hard is negative, and people stop trying. If the probability of reward is high, people try to improve their situation, and those who fail are dissatisfied.

Similarly, Converse (1969) uses a learning model to provide an extremely elegant and exceptionally accurate account (his R^2 is as high as .86) of some "civic culture" differences among the five countries Almond and Verba (1963) studied. According to Converse's account, partisan identification can be learned through participation in democratic institutions. The longer the existence of such institutions, the more stable the partisan attitudes produced. This simple assumption, together with the history of the five nations, can account for the differences in partisan stability that Almond and Verba attributed to differences of "civic culture." Moreover, time can account for the differences between old and new voters, as well as gender differences (women in most of these countries obtained the right to vote relatively recently).

Converse uses learning, not a rational-choice model, to account for these phenomena. However, additional steps can be taken to translate his findings into a rational-choice explanation. Converse claims that time is not a causal agent, though it is a handy indicator of some other process occurring *across* time: learning. If one considers a Bayesian updating process instead of learning,[6] Converse's findings can be explained in terms of rational choice. Older people have stronger priors because they have formed these priors through long experience (a higher number of relevant events). Therefore, revising their attitudes becomes more difficult for them. Younger people have weaker priors, and each new experience is important in forming their beliefs or attitudes. Women in countries that granted women's suffrage only recently are similar to young voters on this account.[7] Consequently, descriptions of historical events or nonrational-choice explanations can be translated into the appropriate rational-choice framework.

---

6. Bayesian updating of information takes place when an individual modifies his previous probability assessments according to Bayes's formula (Skyrms 1986). In this formula, the stronger the priors, the less they are modified by conflicting information.

7. For a similar rational-choice account of the concept of party identification, see Fiorina (1981) and Calvert and McKuen (1985).

To recapitulate, the rational-choice approach is not the only possible one to political phenomena; alternative approaches either study political and social phenomena using actors who do not try to optimize their goal achievement or exclude actors as units of analysis altogether.

Translating relationships across research programs is not always possible. If a translation from some research program to a rational-choice account is possible though not performed, reference will be made to a shortcut (or black box) explanation. If such a translation is impossible, as in the case of Coser, spurious correlation is the result.

II. What the Rational-Choice Approach *Is*

The task here is to derive the implications of the means/ends correspondence with respect to the definition of rationality. I distinguish between two different sets of requirements for rationality: *weak requirements of rationality* and *strong requirements of rationality*. The first assures the internal coherence of preferences and beliefs; the second introduces requirements for external validity (the correspondence of beliefs with reality). Even the weak requirements of rationality are sometimes difficult to meet, which raises the important question of the feasibility and/or fruitfulness of assuming that political actors are indeed rational, a question I answer in Section III.

1. Weak Requirements of Rationality

I discuss the following requirements for rationality: (1) the impossibility of contradictory beliefs or preferences, (2) the impossibility of intransitive preferences, and (3) conformity to the axioms of probability calculus. The first two refer to rational actor behavior under certainty; the third regulates rational actor behavior under risk.

Defending an axiomatic system (in this case, the combination of requirements that define rationality) usually entails demonstrating the plausibility of these requirements (axioms). However, a better argument can be developed by elucidating the undesirable consequences of violating such requirements; the more catastrophic these consequences, the more persuasive the argument. In the demonstrations that follow, I use money to demonstrate undesirable or catastrophic consequences. The advantage of using money to measure the desirability of consequences is the immediate understanding that choices have "objective" consequences for the welfare of individuals. However, one can replicate all my arguments with abstract units of utility (utiles) or some other nonmonetary *numeraire* of satisfaction.

(1) *The impossibility of contradictory beliefs or preferences.* There are two relevant propositions in formal logic: the first claims that the conjunction of a proposition and its negation is a contradiction.[8] The second claims that anything can follow from a false antecedent. If a proposition is a belief, these two laws of logic indicate that anything follows from contradictory beliefs. Therefore, if an actor holds contradictory beliefs, she cannot reason.[9] If a proposition is a preference, the combination of the two laws indicates that anything follows from contradictory preferences. Thus, if an actor holds contradictory preferences, she can choose any option.

Note here that contradiction refers to beliefs or preferences *at a given moment in time.* The impossibility of contradictory beliefs or preferences excludes neither changing beliefs or preferences over time nor holding one preference in one context and a different one in another. It is therefore weaker than the "independence of irrelevant alternatives" axiom in which an actor is assumed to make the same choice between two alternatives whether or not other alternatives exist (Arrow 1951).

(2) *The impossibility of intransitive preferences.* The "transitivity of preferences" axiom states that if an actor prefers alternative a over alternative b, and b over c, she necessarily prefers a over c.[10] It has been demonstrated that one can create a "money pump" (make a lot of money) from a person with intransitive preferences

8. It is, in fact, Aristotle's law of the excluded middle, which can be stated formally as $p \& (-p) = F$, where F stands for "false."

9. Popper (1962) uses this argument to reject dialectical reasoning (which accepts contradictions) as impossible.

10. A similar principle of transitivity in logic assures the possibility of reasoning.

(Davidson, McKinsey, and Suppes 1954). This demonstration is as follows: suppose a person prefers a over b, b over c, and c over a. If she holds a, one could persuade her to exchange it for c provided she pays a fee (say $1). One could also persuade her to exchange c for b for an additional fee (say another dollar). Furthermore, one could persuade her to exchange b for a for an additional fee (say another dollar). Observe that she is in exactly the same situation as before (she holds a); only now she is $3 poorer. In each transaction, she improved her holdings according to her preferences. Because of the intransitivity of her preferences, however, she found herself monetarily worse off. If this money pump continues to function, she could "improve" her situation to the point of starvation.

These two requirements of rationality are part of any rational-choice account because they assure actors' capacity to maximize. The third requirement of weak rationality deals with the objective function that rational actors seek to maximize.

(3) *Conformity to the axioms of probability calculus.* This proposition is the most counterintuitive and most difficult to argue; the proof is presented in the appendix to this chapter. One has to introduce the objective function that a rational agent maximizes. In this book, I assume that rational actors maximize their expected utility, that is, the product of the utility they derive from an event, multiplied by the probability that this event will occur.[11]

The proposition asserts that if a person is willing to make bets in the belief that the probability of winning multiplied by the prize is equal to the probability of losing multiplied by the fee,[12] *and if in*

11. Strictly speaking, there is no reason the decision rule should be part of the definition of rationality. Indeed, one can use different decision rules and derive different predictions. For example, Ferejohn and Fiorina (1974) use the minimax regret criterion to explain why people vote (see *American Political Science Review* [1975] 69:908–60) for an extended discussion generated by their article). Other criteria would be the maximin criterion (Luce and Raiffa 1957) or mixed (Tsebelis 1986) or multiple stage criteria (Levi 1980). However, the overwhelming majority of rational-choice studies assumes that rational actors maximize their expected utility, and this book is no exception.

12. In technical terms, bets with expected utility equal zero. Such bets win $1 if a fair coin comes up heads and lose $1 if it comes up tails, or win $5 if you guess correctly the result of the throw of an unloaded die and pay $1 if you lose. Note that the odds of a fair bet are by far more favorable than the odds that people accept when participating in lotteries or playing in casinos.

her calculations, she does not obey the rules of probability calculus, she will definitely lose money.[13]

For our purposes, this proposition indicates that any individual whose calculus does not conform to the axioms of probability calculus is certain to pay a price (regardless of whether some particular events happen or not) for the inconsistency of her beliefs. For the time being, it does not matter whether an individual's probability estimates correspond to objective frequencies. She may overestimate or underestimate probabilities; she may be optimistic or pessimistic. The only restriction in the proof is that she be willing to accept fair bets, that is, bets with expected utility equal to zero.

In the previous cases, individuals were penalized for deviations from rules of consistency. Some of these rules, of noncontradiction and transitivity, for instance, may seem intuitively pleasing and clear. Others, such as conformity to the axioms of probability calculus, may seem counterintuitive and/or unrealistic. Nevertheless, any deviation from these rules is a deviation from the weak requirements of rationality and will result in a loss of money.

In all the cases, events in the real world were not considered: beliefs had to be (internally) consistent but did not necessarily have to correspond to situations in the real world; in addition, penalties were imposed independently of the state of the world. For example, there is no penalty for the belief in an imminent invasion from Mars as long as the person who has this belief acts consistently with it, that is, prepares herself for the invasion. In order to rule out such possibilities, we must now turn to the external requirements of rationality.

2. Strong Requirements of Rationality

The strong requirements of rationality establish a correspondence between beliefs or behavior and the real world. The following discussion concerns the distinctions among beliefs, probabilities, and

---

13. In this chapter's appendix, I demonstrate that if an individual is willing to make a series of fair bets and her plausibility values do not obey the rules of probability calculus, a Dutch Book can be made against her. The terms *fair bet* and *Dutch Book* are defined in the appendix to this chapter.

strategies, leading to proof of three strong requirements of rationality:

1. Strategies are mutually optimal in equilibrium, or in equilibrium, the players conform to the prescriptions of game theory.
2. Probabilities approximate objective frequencies in equilibrium.
3. Beliefs approximate reality in equilibrium.

It is easier to develop these requirements in reverse order. First, attention must be drawn to the qualifier "in equilibrium," which is present in all three requirements. There are two reasons for this qualification. The first is negative: rational-choice theory cannot describe dynamics; it cannot account for the paths that actors will follow in order to arrive at the prescribed equilibria.[14] The second is positive: equilibrium is defined as a situation from which no actor has an incentive to deviate. Therefore, no matter how equilibrium is achieved, rational actors will remain there.

(1) *Conformity to the prescriptions of game theory.* The Nash equilibrium concept is the fundamental concept of game theory.[15] Players use *mutually optimal strategies* in equilibrium: they have achieved a strategy combination from which no one has an incentive to deviate. According to this definition, there may be more than one equilibrium in a game. The problem then becomes selecting the most reasonable one.[16] If there is more than one reasonable equilibrium, coordination between players becomes a problem. If coordination fails, each player will choose an equilibrium strategy, but these strategies will correspond to different equilibria: the outcome will not be an equilibrium.[17] A player could also deviate

14. In iterated games, one or more equilibrium *paths* can be computed so that the actors change their behavior over time, but they are technically always at equilibrium.

15. Nash (1951). John Nash is one of the founders of game theory.

16. This is the problem of refinements of the Nash equilibrium concept. Several solutions have been proposed: perfect equilibria (Selten 1975), proper equilibria (Myerson 1978), sequential equilibria (Kreps and Wilson 1982b), and stable equilibria (Kohlberg and Mertens 1986). Some of these concepts are discussed in Chapter 3. However, the interested reader should refer to the original articles as well as to Van Damme (1984) for the relationships among these subspecies of Nash equilibria.

17. A simple example is a chicken game, in which both players drive straight toward each other because they believe the opponent will yield, or else both yield

from her equilibrium strategy without being penalized.[18] This deviation, however, may induce other players to change their strategies, either because they are now worse off than in equilibrium or because they can do even better. In both cases, deviation from equilibrium may generate a series of mutual adjustments, leading to the previous equilibrium or to another Nash equilibrium.

Thus, the concept of Nash equilibrium is a necessary (but not sufficient) condition for stability of outcomes. An observer should not expect a situation to be stable if it is out of equilibrium because one of the players will have an incentive to modify her actions. In this sense, the concept of equilibrium is tautological in the context of rational choice.[19] Equilibria are by definition the only combinations of mutually optimal strategies.

(2) *Subjective probabilities should approximate objective frequencies.*[20] This requirement also depends on equilibrium analysis. In game theory, beliefs along the equilibrium path are updated according to Bayes's rule. That is, every player makes the best use of his previous probability assessments and the new information that he gets from the environment. If probability estimates do not approximate objective frequencies, rational actors will be able to improve their results in the long run by revising their probability estimates. Suppose, for example, that one adopts the belief that flipping a coin is a fair bet for heads (it has a one-half probability of landing heads). Suppose also that the coin is

---

because they believe the opponent will drive straight. Chapter 3 explains some of the properties of the game of chicken.

18. She would not be rewarded for such a deviation. In this case, the original position would not have been an equilibrium; she is simply indifferent between the equilibrium strategy and some other strategy. In games with mixed strategy equilibria, indifference between strategies is the rule.

19. This is the dominant position among game theorists. For a proof that only Nash equilibria can be rational solutions to simultaneous games, see Bacharach (1987). For dissenting views concerning sequential games, see Bernheim (1984), Pearce (1984), and particularly Bonanno (1988). The reason for the disagreement is that in sequential games, the calculation of equilibria involves counterfactuals that, by definition, have no truth conditions. For an intermediate position concerning the concept of perfect equilibrium, see Binmore (1987).

20. This assumption is similar to what in the economics literature is known as *rational expectations* (see Muth 1961; Lucas 1982).

biased in such a way that the probability of heads is one-third. After it becomes apparent that losses are more frequent than gains, the player will revise his probability estimates and alter his bets.

(3) *Beliefs should approximate reality.* The argument supporting this requirement is also an equilibrium argument. All beliefs of rational players along an equilibrium path are updated according to Bayes's rule; thus, the actor at any point can chose her optimal strategy given her beliefs. The mutual optimality of the players' strategies (given their beliefs) provides each player information about the beliefs of the opponent. If in the process of the game an actor does not update her information, she may be vulnerable to exploitation by the opponent: the opponent may realize that her situation can be improved given the mistaken beliefs of the first actor. In such a situation, either one of the players would modify her beliefs, or the other would modify her strategy. So such a situation is not an equilibrium.[21]

Consequently, according to the strong requirements of rationality, beliefs and behavior not only have to be consistent, but also have to correspond with the real world (at equilibrium). The penalty for deviations from strong rationality will be a reduced level of welfare.[22]

All arguments concerning both weak and strong rationality are normative. They claim that behavior *should* reflect the prescriptions of expected utility or game theory; otherwise, the actor will pay a price. One could agree with the normative value of these arguments and still not believe that rational choice has any descriptive value. The argument would go as follows: it is true that in an ideally rational world, people should and would behave according to the rational-choice prescriptions, but the real world is very different from such a rational-choice world. In the real world, people are willing to pay the price for their mistakes or for their beliefs; even if real people would like to conform to these prescrip-

21. One more situation exists, characterized by beliefs that have no impact on behavior; thus, there is no reason to modify them. I consider such beliefs innocuous and do not deal with them. The belief in God (without moral imperative supplements) comes as close as possible to such beliefs.

22. I remind the reader that all the proofs can be replicated with utiles instead of money; in this case, one would speak of a reduction in utility instead of welfare.

tions, they are simply incapable of making all the required calculations and computations; calculating Nash equilibria even for simple games is not easy, and the level of complexity increases astronomically when approximating realistic situations.[23]

Is there any reason to believe that the rational-choice approach is, in Keynes's terminology, not only normative but also positive?[24] In other words, are we to believe that real people not only *should* but also *do* behave according to rational-choice requirements? These are the questions I examine in the next section.

III. Is the Rational-Choice Approach Realistic?

A frequent answer to the above question is, "It does not matter; people behave 'as if' they were rational." The full explanation of this particular view is offered in Friedman's seminal article, "The Methodology of Positive Economics." Friedman (1953, 14) claims, "Truly important and significant hypotheses will be found to have 'assumptions' that are *wildly inaccurate* descriptive representations of reality, and, in general, the more significant the theory, the more unrealistic the assumptions (in this sense). ...To be important...a hypothesis must be descriptively false in its assumptions" (emphasis added).

Friedman offers three different examples to support the "F-twist," as economist Paul Samuelson (1963) calls the "as if" thesis. The first concerns expert billiard players who execute their shots "as if" they knew the complicated mathematical formulae describing the optimal path of the balls. The second deals with firms that behave "as if" they were expected utility maximizers. The third concerns the leaves on a tree; Friedman (1953, 19) suggests "the hypothesis that the leaves are positioned as if each leaf deliberately sought to maximize the amount of sunlight it receives."

---

23. The issue of complexity of strategy calculations has only recently become the object of serious investigation. See Kalai and Stanford (1988), Rubinstein (1986), and Abreu (1986).

24. Keynes (1891, 34–35) distinguishes between "A positive science ... a body of systematical knowledge concerning what is; a normative or regulative science ... a body of systematized knowledge discussing criteria of what ought to be."

A similar argument can be made using Hempel's (1964) concept of "potential explanation"—an explanation that is correct if all its premises are true. Nozick (1974) developed this concept in his discussion of the "fundamental potential explanation." He claims that a fundamental potential explanation is important even if it is not true because it reveals important mechanisms that influence the phenomenon under examination. According to these arguments, an explanation may be important even if its premises are not true. So the question of the truth of a theory's assumptions becomes irrelevant.

The "as if" argument claims that the rationality assumption, regardless of its accuracy, is a way to model human behavior. This epistemological position of rationality-as-model is not only partial and unsatisfactory, but also to a large extent responsible for the following situation: on the one hand, several rational-choice explanations use the "as if" argument to justify wildly unrealistic assumptions; on the other hand, empirical scientists mistrust rational-choice explanations for being irrelevant to the real world.

The "rationality-as-model" argument is not satisfactory for the following reason: the assumptions of a theory are, in a trivial sense, also conclusions of the theory. It follows that a scientist who is willing to make the "wildly inaccurate" assumptions Friedman wants him to make admits that "wildly inaccurate" behavior can be generated as a conclusion of his theory. So any scientist who is interested in the realism of the conclusions and explanations of a theory should be concerned with the realism of the assumptions as well. With respect to the rational-choice approach, it is inconsistent to use false assumptions as the basis for explanations after arguing, as I have done, that from false assumptions, anything follows.

I propose a different answer to the realism question. Instead of the concept of rationality as a model of human behavior, I propose the concept of rationality as a subset of human behavior. The change in perspective is important: I do not claim that rational choice can explain every phenomenon and that there is no room for other explanations, but I do claim that rational choice is a better approach to situations in which the actors' identity and goals are established and the rules of the interaction are precise and known to the interacting agents. As the actors' goals become fuzzy, or as the rules of the interaction become more fluid and imprecise, rational-choice explanations will become less applicable. Norton Long (1961, 140–41) provided a similar argument:

Here we deal with the essence of predictability in social affairs. If we know the game being played is baseball and that X is a third baseman, by knowing his position and the game being played we can tell more about X's activities on the field than we could if we examined X as a psychologist or a psychiatrist. If such were not the case, X would belong in the mental ward rather than in a ball park. The behavior of X is not some disembodied rationality but, rather, behavior within an organized group activity that has goals, norms, strategies, and roles that give the very field and ground for rationality. Baseball structures the situation.

I submit that political games (or most of them) structure the situation as well and that the study of political actors under the assumption of rationality is a legitimate approximation of realistic situations, motives, calculations, and behavior. I present five arguments to demonstrate why individuals attempt the calculations described in Section II or why they adopt the behavior prescribed by such calculations or why the aggregate outcome of individual actions can be approximated by such calculations.

Argument 1. Salience of issues and information. Following from the normative properties of the rational-choice approach, people prefer to conform with the behavior prescribed by the theory (otherwise they may have to pay a price). This tendency varies directly with the size of the stakes: for example, candidates try to obtain more information about people's choices in a district with a close race than in a "safe" district; parties spend more resources trying to calculate the consequences of a constitutional modification as opposed to the consequences of a simple law.

Moreover, when information is available, people will be able to approximate the calculations required by rational choice better than when payoffs are not well known or approximate.[25] Indeed, some of the most successful applications of the rational-choice approach concern the institutions, norms, and behaviors of the Congress and bureaucracy of the United States (that is, the study of well-structured situations) (Fenno 1978; Ferejohn 1974;

25. In Chapter 3, I present the folk theorem of repeated games, which asserts that a wide range of outcomes is possible in games with incomplete information.

Fiorina 1974; Hammond and Miller 1987; Miller and Moe 1983; Shepsle and Weingast 1981).

Argument 2. Learning. The normative properties of the rational-choice model suggest that people engaging in repeated activities approximate optimal behavior through trial and error. In fact, subjective probabilities will converge to objective frequencies as additional information becomes available through iteration. Consequently, the final outcome becomes almost indistinguishable from rational-choice calculations. This case is described in one of Friedman's examples: the expert billiard players. The billiard players do not understand the laws of geometrical optics, but they are very sensitive to the implications of such laws for their game. Similarly, voters are able to use retrospective evaluations and vote unsuccessful incumbents out of office in two-party systems even though they may not remember the platforms of the different candidates or be able to discriminate between them (Fiorina 1981; Key 1968).

Learning is not independent of the salience of issues and information. One would expect a correlation between the speed of learning and the salience of the issue, as argument 1 indicates. Moreover, convergence to optimal behavior is faster as the frequency of the decision-making problem increases.

Learning is a conscious activity; it presupposes that the decision maker is able to discover past mistakes. An explanation based on the concept of learning produces the same outcomes as the rational-choice approach but uses much weaker assumptions.

Argument 3. Heterogeneity of individuals. Suppose that instead of adopting the assumption that all individuals can make rational-choice calculations or that all individuals are capable of learning in repeated trials, we make the more realistic assumption that most individuals are not sophisticated, although a small proportion is capable of making these calculations. What will the equilibrium be?

To simplify matters further, assume that a cohort of individuals has to select career paths. Suppose that most have a very simplistic perception of reality and incorrect expectations, but a small percentage is capable of rational-choice calculations. Although the nonsophisticated individuals will make uninformed (and suboptimal) decisions, the most informed will anticipate this behavior and compensate by having their behavior biased in exactly the opposite manner. For example, if there is an excess of doctors, the sophisticated individuals will become engineers or lawyers. So the social outcome will approximate the equilibrium that would prevail if all agents were sophisticated.

This argument has been made by Haltiwanger and Waldman (1985), who prove that equilibria with some sophisticated agents will tend toward equilibria where all agents are sophisticated in the case of "congestion effects," that is, where each agent is worse off the higher the number of other agents who make the same choice as he.[26]

Most economic goods exhibit properties of congestion effects because an increase in demand raises the price and makes additional buyers worse off. I cannot claim that political phenomena demonstrate properties of congestion more frequently than economic ones. However, the number of cases of congestion effects is already large, and in all these cases, an equilibrium with a small number of sophisticated agents is practically indistinguishable from an equilibrium where all agents are sophisticated.

Argument 4. Natural selection. The same behavioral outcomes can be supported, however, by even weaker assumptions. Suppose that there are different "populations" of people defined by their different reactions when faced with the same situation. Furthermore, suppose that when decisions are made and rewards or penalties are distributed, the less successful individuals are eliminated. In the long run, the most successful behaviors are reinforced, and the outcome approximates the optimal choice without any conscious means/ends calculation by those involved.[27] In Friedman's example, firms maximize their expected returns as the result of such an evolutionary process. Similarly, if long-term considerations (as in consistency over time and/or ideology) are excluded, politicians who try to maximize their votes will have a higher

26. The opposite case, in which each agent is better off the more times other agents choose the same behavior as his own (such as buying computer software), is called *synergistic effect*. Haltiwanger and Waldman (1985) prove that in this case, the sophisticated agents imitate the behavior of the nonsophisticated ones, so the latter have a disproportionate effect on the equilibrium.

27. The outcome approximates the optimal choice provided that the population with the optimal choice exists in the beginning of the experiment.

survival rate than those who do not. In the long run, the latter population will be eliminated.

This evolutionary approach adopts the weakest assumptions about individuals' motivations and calculations. In fact, it attributes the entire explanation to environmental factors. For this reason, the explanation is open to the following criticism: a particular behavior is not necessarily optimal because the reason for its natural selection may not have been the behavior under investigation, but some other characteristic. Consequently, evolutionary arguments can be used to support the optimality of behavior only after eliminating alternative explanations.

Argument 5. Statistics. This argument involves the properties of the population mean. Assume the following: only a very small proportion of a population uses rational calculations; only a very small proportion is capable of learning; and evolutionary arguments apply only to a restricted segment of the population. In addition, suppose that the largest remainder of the population makes decisions at random or by some roughly equivalent process. Assume, for example, some are optimists and others pessimists, some are risk prone and others risk averse, and some are influenced positively or negatively in their decisions by opinion leaders (e.g., Ronald Reagan or Jane Fonda).

To make matters more concrete, suppose that rationality is a small but systematic component of any individual, and all other influences are distributed at random. The systematic component has a magnitude x, and the random element is normally distributed with variance s^2. Under these assumptions, each individual of the population will execute a decision in the interval $[x - (2s), x + (2s)]$ 95 percent of the time. If, however, we consider a sample of a million individuals, the average individual will make a decision in the interval $[x - (2s/1000), x + (2s/1000)]$ 95 percent of the time. This can be verified by the statistical properties of the mean: the rational decision assumed to be only a "systematic although very small component" was approximated by the average individual of our sample at a factor of one thousand times that of the random individual. Therefore, the rational-choice analysis can be completely inaccurate concerning a specific individual but very accurate concerning the average individual.[28]

28. The difference between this argument and argument 3 is that here agents are deciding independently; in argument 3, some agents were able to anticipate

There are two possible objections to this statistical argument. First, the problem of aggregation was assumed out. That is, I have assumed aggregation is equivalent to an arithmetic summation. However, as Arrow's (1951) seminal work demonstrates, aggregates (much like societies) can demonstrate properties that are completely divergent from the properties of their constituent parts (individuals). Second, I have arbitrarily equated the systematic component of a decision with rationality. Nonetheless, exactly the same argument can be made if one substitutes any other decision-making rule as a systematic component.

Both objections have merit: if questions of aggregation resembling those described in Arrow's theorem are important, the statistical argument is invalid. Moreover, if any other systematic component is shown to be part of the decision-making process, the decision will gravitate toward this systematic component regardless of its nature. If, for example, people are shown to be risk averse systematically, then risk aversion should be included in the statistical calculations, and the behavior of the aggregate will demonstrate very strong risk aversion.

If the objection concerning the statistical argument in favor of rational choice is essentially correct, why is this argument presented? First, to my knowledge, there are no other claims for *systematic* components of decision making.[29] Second, the reliability of the rational-choice approach does not rest on the statistical

---

the behavior of others and change their choice accordingly. Moreover, in argument 3, the nonsophisticated agents were assumed to be biased; here all agents are normally distributed around some central value.

29. I emphasize the word *systematic* because otherwise theories of national culture or political socialization would be exceptions. There are two directions that systematic dispute of rationality has taken. The first is associated with Tversky and Kahneman (1981), and Kahneman and Tversky (1979), and Kahneman, Slovic, and Tversky (1984) and concerns the framing of decisions. Experiments have indicated important deviations from expected utility-maximizing rules when probabilities are very small or utilities are very large (such as questions of life or death). The second is the satisficing approach associated with Simon (1957), March (1978), and Nelson and Winter (1982), in which people are supposed to choose not the best option among different alternatives but one that is "good enough" or above some threshold of acceptability. The crucial question with respect to this second approach is whether or not people will stick with their choice if some better alternative comes along. In the first case, there is a correspondence between rational choice and satisficing: optimizing refers to the whole set of alternatives, and satisficing refers to a restricted set. But the two methods

argument alone; it rests on all five arguments presented. Each argument is more general but weaker than the previous one. However, taken together, they delineate the range of cases in which the rational-choice approach is legitimate. Validity increases when elites are involved (except when we can use the statistical argument for masses). Validity is practically guaranteed by the existence of small proportions of rational agents in the case of congestion effects. In addition, the results are more likely to be correct in iterated situations in which people learn or are naturally selected than in noniterated games.

To summarize, the rational-choice approach has an indisputable normative appeal. I have also demonstrated that it has a positive value. Contrary to the dominant justification among rational-choice sympathizers, which claims that the validity of the rational-choice approach stems from good predictions, I claim that it is a legitimate approximation of real processes. People will approximate the rational-choice prescriptions when the issues are important, and the degree of approximation will vary with information. Furthermore, there are learning, evolutionary, and statistical reasons why the assumption of optimizing (rational) behavior is appropriate.

The arguments presented here constitute what Musgrave (1981) calls "domain assumptions": necessary conditions for the rational-choice approach. For example, actions taken in noniterative situations by individual decision makers (such as in crisis situations) are not necessarily well suited for rational-choice predictions. Nevertheless, such an approach could have an important heuristic role; it could indicate the realm of possibilities for different actors, demonstrating why certain decisions were or were not made. Rational choice cannot claim to explain all human behavior. Only behavior in situations covered by my five arguments can be the domain of reasonable rational-choice applications.

---

produce the same outcomes when applied to the same set of alternatives (Riker and Ordeshook 1973). In the second case, however, the outcomes are different, and there is no possibility of translation from one research program to the other. These two programs (framing and satisficing) have the advantage of empirical accuracy but have been presented so far as objections to specific claims of the rational-choice program and not as theoretical alternatives.

These arguments further demonstrate that inside the domain of applicability of rational choice, the rationality assumption constitutes a very good approximation of reality. In Musgrave's (1981) terms, for the kinds of cases covered by my five arguments, the rationality assumption is a "negligibility assumption": an assumption that approximates reality so well that it is worth making.

IV. The Advantages of the Rational-Choice Approach

Even if we assume that actors try to do their best under given circumstances, why push the argument to its extreme logical consequences and make the assumption of rationality? Why push the argument to such extremes and then consider it a reasonable approximation of reality? Why try to avoid at all costs explanations including irrational factors or mistakes?

In particular, with respect to the subject of this book, why be surprised when people make suboptimal choices and then try to explain such choices through the use of nested games? Why not simply conclude that the actor is nonrational or that she made a mistake whenever the observer disagrees with her as to the optimal course of action? What is the reason behind this obsession with the rational-choice approach? Why not perform a crucial experiment regarding the rationality assumption and reject rationality if actors choose suboptimally, as is done in experimental psychology? After all, this is the usual treatment of any hypothesis in the social sciences. Put boldly: "why bother inventing epicycles in order to save the rational-choice approach?"[30] There are several reasons.

The social scientist who assumes actors behave rationally makes a reductionist move and at the same time formulates a statement of purpose. She makes a reductionist move because she replaces a series of processes, such as learning, cognition, or mechanisms of social selection, with their outcomes. She does not claim that the actual processes that people use in order to arrive at their rational

---

30. Epicycles were continuously invented to explain the anomalies of the Ptolemaic system. A similar phenomenon occurred before the invention of the theory of relativity. Astronomers were trying to explain anomalies by inventing "hidden planets," that is, planets whose existence would have accounted for the observed anomalies.

decisions are the mathematical formulae used in decision theory or in game theory, but that these formulae lead the scientist in a simple and systematic way to the same outcomes. She makes a statement of purpose because the focus of the study will be on other factors influencing or determining social phenomena.

The rational-choice approach focuses its attention on the *constraints* imposed on rational actors—the institutions of a society. That the rational-choice approach is unconcerned with individuals or actors and focuses its attention on political and social institutions seems paradoxical. The reason for this paradox is simple: individual action is assumed to be optimal adaptation to an institutional environment, and the interaction between individuals is assumed to be an optimal response to each other. Therefore, the prevailing institutions (the rules of the game) determine the behavior of the actors, which in turn produces political or social outcomes.

This approach presents four major advantages over its rivals: theoretical clarity and parsimony, equilibrium analysis, extensive use of deductive reasoning, and interchangeability of individuals.

(1) *Theoretical clarity and parsimony.* Perhaps the most obvious comparative advantage of my approach is in its theoretical clarity and parsimony. Explanations are cast in institutional terms, as opposed to psychological or cognitive process terms. Outcomes are explained as deliberate choices rather than as mistakes. As a consequence, ad hoc explanations are eliminated. If the theoretically predicted behavior does not occur, the notion of mistakes cannot be invoked to explain the actual outcome. Inconsistency between theory and reality is attributed to the inadequacy of the theory rather than to mistaken actors. As a result, the rational-choice approach lends itself to stricter empirical tests than most other theoretical approaches. It also means that the range of potential applications of the theory is limited by its refusal to accept mistakes as explanations of behavior.[31] Although there are phenomena that cannot be explained in rational-choice terms, what rational choice can explain, it explains extremely well because of its parsimony and theoretical clarity.

---

31. In some game theoretic papers, "mistakes" or small perturbations are used as a means to discover stability properties of Nash equilibria (Selten 1975).

(2) *Equilibrium analysis.* One important methodological consequence of the rational-choice approach that is a recurring theme throughout the book is *comparative statics*. Recurring social or political phenomena are considered to be in equilibrium, and the properties of these equilibria are studied and compared. Behavior "in equilibrium" means that the actors involved in a recurring course of action are considered as not having any incentives to deviate from this course. This assumption is a straightforward corollary of the rationality assumption: if a rational actor had an incentive to deviate (that is, improve her condition) from her previous behavior, that behavior was by definition not optimal.

Equilibrium arguments are used in three different ways. First, they are used to discover the optimal behavior of actors. For example, the Socialists in my brief account of Finnish electoral history did not use an equilibrium strategy in the arena of the presidential election; this observation led to the discovery of a nested game in which internal party considerations played an important part. The Socialists' behavior was then explained as an equilibrium (optimal) strategy in this nested game.

Second, equilibrium arguments are used to answer conditional questions and lead to empirically testable predictions. If one of the parameters of the model changes, then one actor may modify her behavior in response to this change; this change in strategy may lead the opponent to a change in her strategy; and this will lead the first actor to further modification; . . . Equilibrium analysis helps us predict the final outcome of this infinite process. Sometimes the prediction is counterintuitive because without the help of mathematical tools the human mind is unable to make the infinite number of calculations required to provide the answer to conditional statements of the form: what would happen if the value of parameter x increased?[32] To use one example, what would happen if the size of the penalty for a particular crime increased? Pure intuition would lead us to expect the frequency of that crime to decline.

32. The legitimate question at this point is, "if game theory is necessary for the analyst to find the counterintuitive solution, then how is it possible for the actors to solve the problem?" The answer provided in Section III is that actual game theoretic calculations are only one of the ways to arrive at an aggregate result. Evolutionary, learning, or statistical averaging arguments would lead the actual actors to the same outcome.

However, game theoretic analysis leads to the conclusion that modifying the criminals' payoffs does not affect their behavior at equilibrium; on the contrary, it affects the behavior of the police.[33]

Third, equilibrium arguments are used to eliminate alternative explanations. Consider theoretical arguments that claim to explain recurring patterns as mistakes, such as the money illusion in Keynesian economics; as habits and rituals, as is often done in cultural anthropology; or as having symbolic meaning, as is often the case in political science. According to rational-choice theory, any explanation that rests on suboptimal behavior is incomplete in the best case and wrong in the worst.

(3) *Deductive reasoning*. The arguments presented in a rational-choice analysis are formal, that is, made according to the rules of mathematics or logic. The advantage of this process is that formal arguments (assuming they are correct) are truth preserving. The conclusions of the models presented carry with them the truth of the assumptions that generated them. In other words, one cannot argue with a theorem (although one can effectively dispute its assumptions). Using provocative terminology, one might say that all rational-choice models are tautological. This tautological quality, far from being trivial, is difficult to achieve. Probably the most important and unambiguous lesson derived from the development of rational-choice "paradoxes" has been that using nonrigorous reasoning frequently leads to wrong conclusions.[34]

Because rational-choice models are tautological, they have two distinctive characteristics. The first is that if a rational-choice model leads to predictions that turn out to be false, the assumptions have to be modified. This is because the methods of derivation of the conclusions are rigorous and truth preserving; there is nothing more in the conclusions than in the assumptions. Logical rigor is not an

33. See Tsebelis (1989), where a simple and plausible game between police and criminals is presented and solved. In equilibrium, modification of a player's payoffs affects the opponent's behavior. In particular, an increase in penalties decreases the frequency of police patrols.

34. People were surprised by Arrow's (1951) possibility theorem because they had never imagined the incompatibility of five restrictions that seemed so trivial and innocuous. McKelvey's (1976) result indicating the omnipresence of majority rule cycles had a similar surprising impact. Many game theoretic results have an important surprise value because the interactions between rational players generates unpredicted outcomes.

exclusive property of formal models, but in these models, the calculations are mechanical and therefore easy to verify. False predictions lead immediately to modification of the assumptions of a model, as opposed to discussions about the logic of the argument.

The second characteristic stemming from the tautological character of rational-choice models is that they permit the cumulation of knowledge. This is because even the models that lead to false predictions are essentially "correct." Once a model has been formulated, it becomes common knowledge that a particular set of assumptions leads to specific results and that a modification of the assumptions or additional assumptions are needed to produce a fit between theory and reality. This is why Arrow's and McKelvey's results stimulated an important stream of research on the importance of institutions.[35]

I believe that the use of deductive reasoning will have an important and lasting influence on political science. Up to now, a common procedure in political science scholarship has been to observe an empirical regularity, then to establish it through statistical methods, and finally to produce a plausible argument consistent with the regularity. The deductive arguments of rational choice have demonstrated conclusively that the last part of this procedure (the presentation of plausible arguments supporting empirical regularities) is not equivalent to theoretical reasoning. Each step of a plausible inductive argument is not completely truth preserving, so by the end of the argument, what is left out may be as or even more important than what was preserved.

(4) *Interchangeability of individuals.* Because the only assumption regarding actors is their rationality, they lack any other characteristic or identity. They are interchangeable.[36] How can a

---

35. See Arrow (1951) and McKelvey (1976). For more recent work on institutions, see Shepsle (1979), Shepsle and Weingast (1984), Riker (1980), and Schwartz (1985).

36. They are interchangeable provided they have the same tastes. As I have already argued, tastes are considered exogenous in rational-choice explanations. One could use the degrees of freedom generated by exogeneity of tastes and provide a "rational-choice" explanation of any phenomenon. For example, one can provide a rational-choice explanation of voting by arguing that there is an intrinsic satisfaction from the act of voting (Riker and Ordeshook 1968). For an argument that this approach is tautological, see Barry (1978). I try to avoid the trap of attributing the essential part of my explanation to tastes by attributing

French Communist be considered interchangeable with an Italian Christian Democrat? What happened to history? What happened to culture? What happened to local tradition? What kind of explanation seems to exclude everything that matters?

It is true that historical, temporal, cultural, racial, or other qualifiers do not enter directly into any rational-choice explanation. However, "The bridge between historical observations and general theory is the substitution of variables for proper names of social systems in the course of comparative research" (Przeworski and Teune 1970, 25). The rational-choice research program is not the only one that tries to replace ethnic or racial characteristics or behaviors with the goals of the actors or the institutions that produce them. If Italians are cynical, Germans obedient, and Mexicans distrustful of government, as *Civic Culture* indicates, it is not because they are Italians, Germans, or Mexicans. Section III gave examples of how some of Almond and Verba's findings can be explained in terms of existing institutions and the assumption of rationality.

I now focus on one particular interchangeable actor: the reader. In the rational-choice approach, outcomes are explained as the optimal choices of actors in a given situation. A successful rational-choice explanation describes the prevailing institutions and context in which the actor operates, persuading the reader that she would have made the same choice if placed in the same situation.

This is the notion of *Verstehen* (understanding), central to the Weberian account of social science. Weber distinguishes two kinds of *Verstehen*: direct observational understanding and explanatory understanding that "seeks to grasp the 'motivation' or the final cause of behavior by 'placing the act in an intelligible and more inclusive context of meaning'" (Dallmayr and McCarthy 1977, 21).

"standard" tastes to my actors. One such standard taste for political actors is reelection because it is a necessary condition for achieving any other political goal. For example, in Chapter 5, Labour party activists are considered to have ideological preferences, but not to the point of sacrificing their party's candidate; Belgian elites have preferences over outcomes, but not to the point that it will cost them reelection; and French parties want to improve their electoral position without hurting the electoral chances of their coalition.

The concept of explanatory understanding was rejected by the positivist tradition in the social sciences because it presupposes some "empathetic identification," "personal experience," or "introspective capacity" and therefore is a subjective process or method (Abel 1948; Rudner 1966). Moreover, *Verstehen* explanations are only potential explanations, establishing the possibility of certain relations or connections, and cannot be empirically tested because the empathetically explained phenomenon cannot be replicated. For similar reasons, the concept of *Verstehen* was embraced by the hermeneutic tradition of social science (Taylor 1965).

An immediate consequence of my understanding of *Verstehen* is that both the positivist rejection and the hermeneutic appropriation of the concept were hasty and misplaced. The interchangeability of individuals, or *Verstehen*, as applied in this book and in other rational-choice approaches is immune to positivist criticisms: understanding does not depend on any subjective psychological capacity of empathy but on the application of strict rules of optimal behavior under constraints.

Moreover, the criticism of testability is based on the epistemological position of symmetry between explanation and prediction (Hempel 1964). This particular epistemological position has been rejected by most philosophers of science (Scriven 1962). It is possible to predict without explaining (the obvious examples include weather reports and economic forecasts) or to explain a posteriori things that could not have been predicted a priori (preemptive wars). Although unique social phenomena can occasionally be understood yet not replicated (and therefore the explanations are not testable and do not lead to predictions), such explanations are not less scientific than testable propositions that lead to predictions. Therefore, the interchangeability of individuals or actors and the interchangeability with the reader is not a liability of the rational-choice approach. Instead, it is its strength: it constitutes a conscious effort to apply standards of scientific explanation in the social sciences.[37]

For all these reasons, the hypothesis of rationality is not placed

37. For a similar analysis of the concept of *Verstehen*, see Scharpf and Ryll (1988).

at the same level as other hypotheses in a rational-choice explanation. In fact, it is the core concept of the whole rational-choice research program. Lakatos (1970) calls such core concepts "negative heuristic," indicating that as long as the research program is alive, they cannot be modified. In order to achieve this goal, every research program forms a protective belt around these concepts, a series of auxiliary hypotheses that Lakatos calls "positive heuristic." These auxiliary hypotheses must be modified if there are inconsistencies between the predictions of a theory and reality.

In fact, Lakatos claims we can never perform a crucial experiment on an isolated hypothesis—we always test joint hypotheses. If the results of these tests are negative, at least one of the elementary hypotheses that form the joint hypothesis must be rejected, but we do not necessarily know which one. At this point, some of the positive heuristic concepts or hypotheses must be sacrificed to save the negative heuristic.

All the models presented in this book make two kinds of assumptions: rationality of the actors in the sense defined in this chapter and certain institutional structures. These assumptions jointly lead to explanations or predictions. If these explanations or predictions turn out to be false, then some of the initial assumptions will have to be modified. Lakatos's negative heuristic concept indicates that the appropriate modifications inside the rational-choice research program are the ones concerning the description of institutional structures, *not the assumption of rationality*. Because this assumption is the core of rational choice and of economics and because it constitutes the implicit basis of most mainstream political science, there is no reason to dispute it every time an anomaly is presented.

This book elaborates the nested games concept in order to account for puzzles and anomalies not as failures of rationality, but as indications of the systematic impact of contextual or institutional factors. When these factors are taken into account, the actors' behavior becomes intelligible.

To recapitulate, the rational-choice approach assumes the individual's behavior is an optimal response to the conditions of her environment and to the behavior of other actors. A successful rational-choice explanation describes prevailing institutions and existing contexts, persuading the reader that the action undertaken was optimal and that she would have adopted the same course of action in the same situation.

It is time now to apply the principle of rationality to some concrete cases. First, however, in order to make specific applications, some elementary game theoretic notions are necessary. I provide these in Chapter 3.

Appendix to Chapter 2

For reasons of simplicity, all arguments are made using money to express utilities. However, every argument can be replicated with *utiles*, that is, an abstract and linear *numeraire* of utility. Arrow (1965) and Pratt (1964) have defined risk aversion as the degree of concavity of a utility function, so the use of utiles instead of money includes standard economic definitions of risk proneness or risk aversion in my account.[38]

A different but equivalent version of the axioms of Kolmogorov is used to prove that conformity to the axioms of probability calculus is a weak requirement of rationality.[39] These axioms are the following:

A1. No probability is less than zero. Formally, $P(i) > = 0$

A2. The probability of a sure event is one. Formally, $P(I) = 1$

A3. If i and j are two mutually exclusive events, then $P(i \text{ or } j) = P(i) + P(j)$

I demonstrate the following proposition: *if a person is willing to make a series of fair bets and her plausibility values do not obey the axioms of probability calculus, a Dutch Book can be made against her.*

For the proof, two definitions are needed. First, a *fair bet* is defined as a bet with the following property: if one is willing to bet an amount of money (say a) and receive an amount of money (say

---

38. However, accounts of attitudes toward risk that are defined by non-linearities in the probability part of individual calculations (Chew 1983; Edwards 1954; Fishburn 1983; Kahneman and Tversky 1979; Karmarkar 1978; Machina 1982) violate the assumptions of my account of rationality.

39. Andrey Nikolaevich Kolmogorov, a Russian mathematician, founded axiomatic probability calculus.

b) if the bet is won, then the ratio a/(a + b) (the betting quotient) is equal to the probability of winning. A fair bet is by definition one with a betting quotient equal to the probability of winning, or, equivalently, a fair bet is a bet with an expected utility equal to zero.

A *Dutch Book* has been made against someone if the amount of the bet, *no matter what happens* in the real world, is lost.

The proof proceeds in three steps and demonstrates the consequences of violating each axiom (Skyrms 1986).

1. Violation of Axiom A1

Suppose an individual assigns a negative plausibility to an event. It then follows that she considers a bet with negative betting quotient [a/(a + b)] as fair. Therefore, she will be willing to accept a bet with negative winnings (a) and positive losses (b).[40]

Example. If an individual assigns a plausibility of $-.2$ to an event e, she will be willing to accept a bet in which she wins -60 if e is true and pays 10 if e is false. Indeed, such a bet has a betting quotient of $-.2$, which is equal to the plausibility of e. The outcome of this bet will be -10 if e turns out to be false and -60 if e turns out to be true.

2. Violation of Axiom A2

There are two possible cases: a sure event (I) can be considered as having plausibility either greater or less than 1. First, suppose a plausibility greater than 1 is assigned to I. In this case, a bet that has a betting quotient greater than 1 is considered fair. Therefore, the bet will be accepted with negative winnings (a) and positive losses (b).

Example. If an individual assigns a plausibility of 1.5 to a sure event e, she will be willing to win -10 if e is true and pay 30 if e is false. Indeed, this bet has a betting quotient $30/(-10 + 30) = 1.5$.

Now suppose that an individual assigns a plausibility of less than 1 to a sure event I. In this case, she is willing to bet against e

---

40. Note that the betting quotient is negative as long as the absolute value of the winnings is smaller than the losses.

at a particular betting quotient. That is, she will accept a bet with negative winnings and positive losses: a regrettable action.

Example. If an individual assigns plausibility .75 to a sure event e, she will be willing to win −75 if e is true and pay −25 (that is, receive 25) if e is false. Indeed, this bet has a betting quotient −75/(−25 − 75) = .75.

3. *Violation of Axiom A3*

Again, there are two possible cases: an individual assigns a plausibility value to a composite (i or j) event either greater or smaller than the sum of plausibilities of elementary (mutually exclusive) events i and j.

Suppose first that P(i or j) < P(i) + P(j). Because the individual assigns plausibility P(i) to event i, she is willing to accept a bet that pays $1 - P(i)$ if i occurs and loses P(i) if i does not occur. Indeed, the betting quotient is P(i). Similarly, if the plausibility of j is P(j), she is willing to accept a bet that pays $1 - P(j)$ if j occurs and pays P(j) if j does not occur. Finally, if the plausibility of (i or j) is P(i or j), she is willing to accept a bet that pays P(i or j) if (i or j) does not occur and pays $1 - P(i$ or j) if (i or j) occurs.[41] Let us now see what happens if all three bets are accepted.

There are three possible outcomes: i occurs and j does not, j occurs and i does not, or neither occurs.

If i occurs and j does not, $1 - P(i)$ is received for the correct guess concerning i, P(j) is paid for the wrong guess concerning j, and $1 - P(i$ or j) is paid for the wrong guess concerning (i or j). The net outcome of these transactions is $1 - P(i) - P(j) - (1 - P(i$ or j)). After simplification, the net outcome is $P(i$ or j) $- P(i) - P(j)$.

If j occurs and i does not, the individual pays P(i) for the wrong guess concerning i, receives $1 - P(j)$ for the correct guess concerning j, and pays $1 - P(i$ or j) for the wrong guess concerning (i or j). The net outcome of these transactions is again $P(i$ or j) $- P(i) - P(j)$.

If neither i nor j occurs, the individual pays P(i) and P(j) for the wrong guesses concerning i and j and receives P(i or j) for the

41. Readers can easily verify that this bet is also fair.

correct guess concerning (i or j). The net outcome is again P(i or j) − P(i) − P(j).

Thus, no matter what the situation in the real world is, the individual receives a net outcome of P(i or j) − P(i) − P(j) from her series of bets. However, this net outcome, by assumption, is *negative*.

Suppose P(i or j) > P(i) + P(j). In the previous case, the individual was willing to accept what were considered fair bets *for* i, *for* j, and *against* (i or j). In a similar way, she will be willing to bet *against* i, *against* j, and *for* (i or j).[42] By replicating the steps of the previous argument, we can show that the net outcome will be P(i) + P(j) − P(i or j) this particular time. The net outcome, by definition, is *negative*.

QED.

42. Readers are reminded that, according to the definition of a fair bet, one can accept a bet for or against an event when the bet is fair.

Chapter Three

Two-Person Games with Variable Payoffs

In Chapter 1, I said that I would represent games in multiple arenas as games with variable payoffs, games in which the payoffs of the game in the principal arena are influenced by the prevailing conditions in another arena. This chapter serves two purposes: to explain why games in multiple arenas can be represented by games with variable payoffs and to introduce the reader to the study of games with variable payoffs.

Section I analyzes the relation between ordinary game theoretic concepts such as equilibria and subgames, on the one hand, and games in multiple arenas, on the other, and explains the use of the concept of nested games in analyzing political situations. Section II examines four different two-person games (prisoners' dilemma, assurance or stag hunt, chicken, and deadlock) in an attempt to familiarize readers with fundamental concepts of game theory. The section introduces one-shot games, their solutions (their equilibria), and the kinds of visual representations used throughout the book. Section III deals with the same four games when contingent or correlated strategies are possible. I show that in this case, the equilibria of the four games are multiplied. However, when contingent strategies are possible, the likelihood of different equilibria varies with the magnitude of each player's payoffs. I argue that for all four games, the likelihood of cooperation increases when the payoffs from cooperation increase and when the payoffs for defection decrease. This is the major finding and will be used repeatedly throughout the rest of the book. Section IV shows that exactly the same relationship between payoffs and cooperation holds in the case of iterated games.

One note on the method of presentation. Section III contains results used repeatedly in subsequent chapters. To facilitate comprehension, I present the argument in Section III and prove it formally in the two appendices to this chapter. I have chosen this format of presentation in order to make the rest of the book accessible to nontechnically inclined readers. The same principle of clarity of exposition applies to the appendices: Appendix A is simpler and deals with the prisoners' dilemma game, which has attracted so much attention in the literature; Appendix B generalizes to the remaining three games. Both appendices contain not only technical material, but also important points that cannot be made without reference to equations or figures and that will be extremely useful to people who, besides reading this book, would like to apply the same line of reasoning to other problems.

I. Games, Subgames, and Nested Games

Why and how would the payoffs of a game vary? To answer this question, I introduce a simple three-person game that demonstrates the logic of the arguments that follow.

Figure 3.1 represents such a three-person game tree, where one player (in this case, player 1) moves first, and the other two move simultaneously. The sequence of moves is indicated by the convention that previous choices (appearing higher in the game tree) are known by all subsequent players and that simultaneous moves are indicated by the dotted lines, which are called *information sets*. An information set indicates that the player who chooses is not able to discriminate between the nodes connected by the information set and consequently does not know the move of the player that precedes him.

In Figure 3.1, player 1 moves first, chooses whether to go left or right, and by this choice determines whether the other two players will play the right- or the left-hand side of the game tree. The figure also indicates that the other two players move simultaneous-

Nested Games

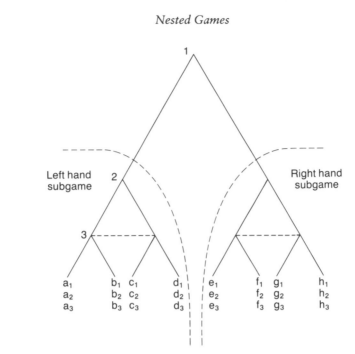

Figure 3.1 Three-person game.

ly because player 3 has to move without knowing (because of the information set) player 2's choice. However, both know whether player 1 has chosen the right or the left side of the tree.

The choice of one strategy by each player leads to a jointly determined outcome that entails a payoff for each player. These payoffs may be positive or negative. The payoffs are indexed by player. There are $2^3 = 8$ possible outcomes, and each player's payoffs are presented as a column at each of the final nodes of the game tree. By convention, player 1's payoff is written first (at the top of each column), player 2's payoff is written second (in the middle), and player 3's payoff is written third (at the bottom). One additional convention I adopt in this book is that odd numbered players are female and even numbered players are male. In particular, for the two-person games that constitute the major part of the games in this book, the first player is female and her opponent is male.

Because all players are rational, they will make their choices according to the rules of game theory; therefore, they will choose mutually optimal (that is, equilibrium) strategies, as noted in Chapter 2. In particular, players 2 and 3 will make mutually optimal choices corresponding to player 1's choice.[1] In her turn, player 1 can choose her optimal strategy for two reasons. First, she can anticipate her opponents' reactions in each one of her moves (because she knows their payoffs and assumes rationality on their part). Second, she knows her own payoffs. Therefore, she will choose the strategy that maximizes her payoffs, given that the other two players will play their equilibrium strategies.

Game theory has developed a very important concept that simplifies such situations: the concept of subgames (Ordeshook 1986, 139). For our purposes, a *subgame* is a game between two or more players that can be completely isolated from the games around it and can be solved (that is, the equilibria can be computed) on its own.

The game in Figure 3.1 has two subgames: the right and the left sides of the tree, which are defined by player 1's choice. In both subgames, the equilibrium strategies can be calculated, so player 1 knows (because of the rationality assumption) how players 2 and 3 will respond to her initial choice and can choose the option that will maximize her payoffs.

Finally, the observer knowing all three players' payoffs can make the same calculations, solve the game, and predict its outcome. This is the game theoretic approach to the situation presented in Figure 3.1.

Consider now the following complications: suppose player 1 is "nature," that is, some kind of lottery that decides which of the two subgames the other two players (that is, players 2 and 3) will play.[2] Or consider that "player 1" may be a shorthand expression for a series of players who interact with one another and jointly select one strategy or the other. Although each player may be rational, the final selection need not be the best they can do *collectively*. Finally, and more congruently with what I have argued so far, it is possible that player 1 is involved in games in multiple arenas and that her payoffs depend on the situations in all these

1. In Section II, I explain how these mutually optimal choices are calculated.
2. This conceptualization of random events is standard in game theory, particularly in games with incomplete information (Harsanyi 1967–68).

arenas. In all these cases, the observer does not know player 1's payoffs.

Ignorance of player 1's payoffs makes it impossible for the observer to analyze the situation and find out what player 1 will do; consequently, it is impossible for the observer to solve the game. However, it is still possible to solve both of the subgames and say that, regardless of whether player 1 chooses the right- or the left-hand side of the tree, the other two players will respond by choosing mutually optimal strategies (in the corresponding subgame). The equilibrium strategies for the left-hand subgame may or may not be behaviorally the same as the equilibrium strategies in the right-hand subgame. For example, it is possible that under a certain set of conditions (that is, if player 1 chooses the right-hand subgame), player 2's best option is to be aggressive, but under another set of conditions (that is, if player 1 chooses the left-hand side subgame), his best option is to be conciliatory.

If the equilibrium strategies in both subgames are behaviorally the same, knowing player 1's choice is not necessary to understand what players 2 and 3 will do. If, however, the equilibrium strategies in the two subgames are behaviorally different, the observer cannot predict the behavior of players 2 and 3. And if the observer believes that players 2 and 3 are playing the right-hand subgame, while in fact they are playing the left-hand subgame, then he will make wrong predictions about their behavior and will be surprised by the actual way they behave. This captures the element of surprise we will encounter in each of the empirical studies in this book. Political leaders or followers or other political parties will disturb the game we study; the disturbance is sufficient for the principal actors to modify their mutually optimal behavior (their equilibrium strategies).

Why should we use this complicated conceptualization where some of the players have payoffs known to the observer and some do not in order to study political situations? Why not model the situation as a game with an appropriate number of rational players so that every relevant aspect of a political situation would be included? There are two reasons. First, the situation may actually be as I have described it, that is, some actor may be a social aggregate or nature or involved in another arena. Second, even if a rep-

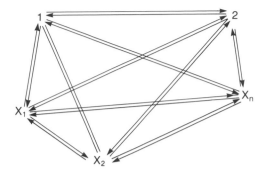

Figure 3.2 Graphic representation of n-person game.

resentation of all relevant aspects of a political situation were possible, such complicated games are usually intractable.[3]

Figure 3.2 illustrates such a situation. It represents a game with n players. The two-sided arrows indicate the interactions among the different players of the game. Suppose the appropriate specifications of these arrows lead to an accurate representation of the n-person game. Given that the solution of this game is impossible, how can we study the situation? The game theoretic answer is the concept of subgames.

Figure 3.3 presents one subgame in a schematic way.[4] The double-sided arrows connecting players 1 and 2 with the rest of the players in Figure 3.2 have been eliminated. In other words, in a subgame, by definition, there is no influence exercised by contextual factors.

3. For a remarkable exception, see Austen-Smith and Banks (1988), where a multistage game between voters, parties, and coalitions is studied in detail. The article is extremely innovative because it deals with coalition formation as a non-cooperative game, examines the legislative and the electoral arenas at the same time, and contrary to prevailing beliefs (Duverger's laws), demonstrates the possibility of strategic voting in proportional electoral systems. However, their approach is analytically complicated despite the fact that, as they acknowledge, they trade off "generality in favor of analytic tractability."

4. A game may have several subgames, and each subgame may in its turn have several subgames. Moreover, two actors may be involved in several subgames. Figure 3.2 is not designed to present these complications. For clarity of exposition, here I present only one subgame.

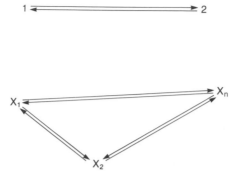

Figure 3.3 Graphic representation of two-person subgame.

If one could study the complicated game in Figure 3.2, or if this game could be reduced to subgames, as in Figure 3.3, there would be no need to study games in multiple arenas. Unfortunately, most of the time, the game presented in Figure 3.2, although a complete and accurate representation of reality, is intractable; and most of the time, the subgame presented in Figure 3.3, although simple and tractable, is an inaccurate or unrealistic representation of social or political situations. Compared to these inadequate representations, games in multiple arenas can be conceptualized as games in which the situation prevailing in other arenas determines the payoffs of the players in the principal arena.

Consider the election of the Finnish president presented at the beginning of Chapter 1. What we call "the game in the parliamentary arena" in ordinary language is a series of possible subgames in game theoretic terms. Which one of these subgames will actually be played out by the parliamentary actors depends on the actions of other actors (the activists and voters of each party). If the activists and voters choose to remain passive observers of the parliamentary arena (as the Communists did), or if they choose to interfere or to threaten to interfere (as the Socialists did), one particular subgame is selected and played out. So games in multiple arenas are a means to study all possible subgames that depend on contextual factors.

Another way to express the same idea is through the concept of *externalities*, that is, consequences to third parties from the interaction between two other parties. If externalities are negligible, then the interaction between two actors can be studied in isolation, and predictions about each actor's behavior can be accurate; this is the case for Figure 3.3. If, however, externalities are important, and if third parties react to the game between the two players, then these externalities have to be studied. One way of studying externalities is to consider them explicitly and construct a complete model of the interaction among all interested parties; this is the case in Figure 3.2. Another way of studying the interaction between two players with externalities is by focusing on the

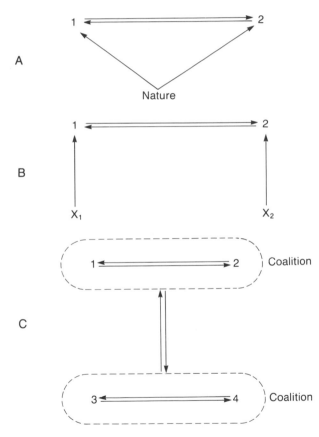

Figure 3.4A Graphic representation of nested game in Chapter 5.
Figure 3.4B Graphic representation of nested game in Chapter 6.
Figure 3.4C Graphic representation of nested game in Chapter 7.

impact of the reactions of third parties to the interaction of the main players; this is the nested games approach. Figure 3.4 gives graphic representations of such nested games.

In Figure 3.4A, all the other players of Figure 3.2 are replaced by one player ("nature"), who determines the payoffs of the two players in the principal arena. Figure 3.4A essentially represents a comparative statics problem: outside conditions change the payoffs of the actors, and the equilibria of two different games have to be calculated and compared (see Chapter 2). A concrete application of such a game in multiple arenas is offered in Chapter 5, where the payoffs of Labour activists and MPs will be decided by an outside factor: the competitive game between parties at the constituency level. The two players in the principal arena will have different payoffs and therefore will select different strategies according to whether their constituency is marginal or safe.

In Figure 3.4B, all the additional players of Figure 3.2 have been reduced to two players; player X_1's actions affect player 1's payoffs, and player X_2's actions affect player 2's payoffs. In Chapter 6, I provide a specific example of such a game. Belgian elites will play one game among themselves in the parliamentary arena, and each one's payoffs will depend upon the interaction with its followers and the situation in the electoral arena.

Figure 3.4C presents a more complicated situation in which the game between two partners in a coalition is affected by the competition between their coalition and a rival. Chapter 7 applies this representation of games in multiple arenas to the study of French electoral coalitions.

Readers can extend the figures on their own. One can think of cases in which only one player's payoffs are affected by some outside event or player. In this case, the graphic representation resembles a gamma rather than a pi (as in Figure 3.4B). One can think of one player's payoffs depending on two outside events or players, and so on. Finally, one can start applying the same framework to three-person games.

To summarize, games in multiple arenas introduce politi al context to game theoretic problems. The payoffs of the players in the principal arena vary according to the situation prevailing in other arenas or the moves made by players in these arenas. The use of games in multiple arenas is in studying situations in which political

TABLE 3.1.   *Payoff matrices and definitions of four games.*

	Cooperate	*Defect*
Cooperate	$R_1 \, R_2$	$S_1 \, T_2$
Defect	$T_1 \, S_2$	$P_1 \, P_2$

Ti > Ri > Pi > Si: Prisoners' dilemma
Ti > Pi > Ri > Si: Deadlock
Ti > Ri > Si > Pi: Chicken
Ri > Ti > Pi > Si: Assurance

context is important and the situation is so complicated that reference to exogenous factors is required.

II. Some Simple Two-by-Two Games

Readers by now know that the payoffs of games may vary according to which subgame the actors actually play. What remains to be seen is how these variations in payoffs affect the behavior of the players in the principal arena and what the consequences will be of these variations for the equilibria (that is, the empirical predictions) of different games. This section studies four simple one-shot two-by-two games: prisoners' dilemma, assurance, chicken, and deadlock games.

Table 3.1 presents the generic payoff matrix of these four games. Each player has a choice between two strategies. To simplify matters, I assume that in all four games, the players have a choice between the same two strategies: cooperate or defect. The choice of one strategy by each player leads to a jointly determined outcome that entails a payoff for each player. For reasons of simplicity, the payoffs in all games are symbolized by the same letters (T, R, P, and S); what differs is the order of these payoffs in each game. In the case of mutual cooperation, each player receives a reward R_i (R is a mnemonic for reward, and i is the index of the player, that is, it can be either 1 or 2). In the case of mutual defection, each player receives a penalty P_i (P stands for penalty). If one cooperates while the other defects, the cooperative player receives the sucker's payoff S_i (S for sucker), and the defecting player receives the temptation payoff T_i (T for temptation).

The presentation begins with the most frequent and familiar two-person game: the prisoners' dilemma. The game was invented by Flood (1952) and given its name by Tucker (1950), who invented the supporting story.[5] The game has been used to study the problem of the emergence of cooperation among rational self-interested agents. The emergence of cooperation is important for political philosophy (Taylor 1976), international politics (security dilemma [Jervis 1978] and disarmament [Rapoport 1960]), and political economy (cartels [Laver 1977], concertation [Lange 1984], the study of economic and social exchange [Calvert 1985], collective action [Axelrod 1983; Hardin 1971], public goods [Head 1972; Samuelson 1954], and markets [Hardin 1982]).

The prisoners' dilemma game has two characteristics. First, defection is the dominant strategy for each player. *Dominant* is the technical term used to indicate that following this strategy leaves each player better off *no matter what the opponent does*. Thus, defection is unconditionally the best strategy for each player. Second, by choosing the dominant strategy and defecting, both players find themselves in a suboptimal outcome, that is, they find themselves worse off than if they had chosen the cooperative strategy.

The relationship between the different payoffs for prisoners' dilemma is the following:

$$T_i > R_i > P_i > S_i \qquad\qquad (3.1)$$

One can verify that under condition (3.1), each player is better off when he or she defects no matter what the other does (dominance). Indeed, if the opponent chooses to cooperate, defection provides a higher payoff (T_i) than cooperation (R_i). Similarly, if the opponent chooses to defect, defection is still better (P_i) than cooperation (S_i). As a result, both players find themselves with a suboptimal outcome because each receives P instead of R.

---

5. The story goes as follows: two prisoners suspected of a serious crime are held in different cells, and each is offered the following deal by the district attorney: "If you confess and the other prisoner does not confess, you will be let free; if the other prisoner confesses, too, you will receive a moderate sentence. If neither of you confesses, you will receive a smaller sentence than if you both confess; if the other confesses but you do not, you will receive the maximum sentence."

This relationship of dominance between the two available strategies provides a strong incentive for each player to defect. In fact, not only expected utility maximization but a wide range of decision rules (maximin criterion and minimax regret criterion, to mention only two) require that if there is a choice between a dominant and a dominated strategy, the dominant one is better.

Here lies the dilemma of the prisoners: they would prefer to be able to communicate and arrange their defense in such a way that they would both be better off. However, in the absence of communication, each can either choose the dominant strategy, which will make both of them worse off, or choose the dominated strategy and receive the sucker's payoff S_i.

If for each player we reverse the order of the penalty P_i and reward R_i, then a different game is generated: deadlock. The ordering of the different payoffs in a deadlock game is as follows:

$$T_i > P_i > R_i > S_i \qquad\qquad (3.2)$$

Deadlock has been used extensively in the international relations literature (Oye 1986; Snyder and Diesing 1977). Deadlock shares with the prisoners' dilemma the feature of a dominant strategy (defection). It differs, however, because defection does not produce a suboptimal outcome: both players are better off with mutual defection than with mutual cooperation.

The remaining two games, chicken and assurance, feature the common characteristic of having no dominant strategy. In the game of chicken, mutual defection is the worst possible outcome for both players. Condition (3.3) represents the payoffs for the game of chicken.

$$T_i > R_i > S_i > P_i \qquad\qquad (3.3)$$

The fear of arriving at this worst possible outcome, in which the payoff for each player is P_i, may (under conditions to be specified subsequently) lead both players to cooperate.

In the assurance game, mutual cooperation is the preferred outcome. The payoffs of the game follow condition (3.4).

$$R_i > T_i > P_i > S_i \qquad\qquad (3.4)$$

As noted in Chapter 2, rationality implies that players conform to the prescriptions of game theory, choosing mutually optimal strategies. I called these pairs of strategies (Nash) equilibrium strategies, stating further that their choice will lead to equilibrium outcomes. Equilibria are stable outcomes because no player has the incentive to deviate in strategy if the opponent does not change strategy. What are the equilibria of these four games?

In prisoners' dilemma and deadlock, the players have the dominant strategy of defection. Therefore, the outcome in both games is the intersection of the "defect" strategies, and the payoffs are (P_1, P_2). The remaining two games have two equilibria each. In chicken, if player 1 (the row player) chooses to defect, player 2 (the column player) is better off cooperating, and if player 2 cooperates, player 1 is better off defecting. Thus, defection from player 1 and cooperation from player 2 are mutually optimal strategies; once the players choose this combination, they have no unilateral incentive to deviate. For similar reasons, cooperation from player 1 and defection from player 2 is also an equilibrium. The payoffs in these two equilibria are (T_1, S_2) and (S_1, T_2). In the assurance game, if one player chooses to cooperate, the other is better off cooperating, but if one player chooses to defect, the other is better off defecting as well. This reasoning indicates that the outcomes (R_1, R_2) and (P_1, P_2) are the two equilibria of the game.

In games with multiple equilibria, it is possible that one of them will be selected by both players. In the assurance game, for example, both (R_1, R_2) and (P_1, P_2) are equilibria, but the first makes both players better off. Therefore, each player can anticipate that both she and her opponent will choose to cooperate, and the outcome will be (R_1, R_2). It is also possible that such a selection between equilibria cannot be made. In the game of chicken, both (T_1, S_2) and (S_1, T_2) are equilibria, but there is no obvious way of choosing between them. The row player prefers the first; the column player prefers the second. This divergence of preferences generates a problem of coordination between the two players. If, somehow, one player makes it clear that she will not cooperate, the other will acquiesce and cooperate. If communication is not possible or fails, both players may defect (because each tried to force her own preferred equilibrium upon the other), or they may both cooperate (because each was afraid of the other's defection). Thus, the multiplicity of equilibria is a source of instability of out-

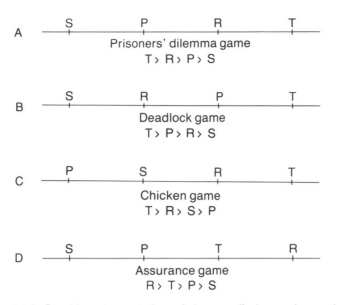

Figure 3.5A Graphic representation of the payoffs in a prisoners' dilemma game.

Figure 3.5B Graphic representation of the payoffs in a deadlock game.

Figure 3.5C Graphic representation of the payoffs in a chicken game.

Figure 3.5D Graphic representation of the payoffs in an assurance game.

comes. Three of the four games present either unique equilibria or the possibility of selection between equilibria. In the game of chicken, however, selection between equilibria is problematic.

One additional source of instability stems from the question of whether the outcome was optimal for both players or could be improved upon: the question of Pareto optimality. Technically, an outcome is called *Pareto optimal* if it is impossible to improve one player's payoff without reducing another's. An outcome that is not Pareto optimal presents the following source of instability: the players know that if they get together, they can improve the payoffs for some (or for all) of them. Because communication is prohibited, however, such an agreement is not possible. Figures 3.5 and 3.6 represent the relationship between the payoffs of the different games and the problem of Pareto optimality.

Figure 3.5 is a graphic representation of one player's payoffs in each game. Figure 3.5A shows the payoffs of a prisoners' dilemma

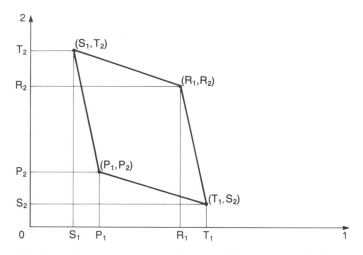

Figure 3.6 Graphic representation of the payoffs in a prisoners' dilemma game on a plane.

along one axis. Figure 3.5B shows that deadlock can be generated if the order of P_i and R_i is reversed. Figure 3.5C demonstrates that a reversal of P_i and S_i transforms a prisoners' dilemma game into a game of chicken. Figure 3.5D demonstrates that a reversal of R_i and T_i produces an assurance game. If we represent the payoffs of each player of a prisoners' dilemma game along each one of the two axes of Figure 3.6, we get a graph of the game.

Figure 3.6 represents a prisoners' dilemma game because the order of each player's payoffs is similar to Figure 3.5A. In Chapter 6, I use a similar representation of the chicken game to analyze how Belgian institutions produce Pareto optimal outcomes. Readers can verify that there are three points representing Pareto optimal outcomes in this figure: (T_1, S_2), (R_1, R_2), and (S_1, T_2). For each outcome, it is impossible to improve one player's payoff without hurting the other. Conversely, there is one outcome that, if selected, produces unsatisfactory results for both players: (P_1, P_2). This is precisely the equilibrium of the prisoners' dilemma game. Therefore, when the players of a prisoners' dilemma game choose their dominant strategies, they produce an outcome that is not Pareto optimal. Note, however, that in order to actually improve upon this outcome, they need to violate the rules of the

TABLE 3.2. *Game theoretic properties of the four games.*

	Equilibria	Is Eq. Pareto Optimal?
Prisoners' dilemma	$P_1 P_2$	No
Deadlock	$P_1 P_2$	Yes
Chicken	$S_1 T_2, T_1 S_2$	Yes
Assurance	$R_1 R_2, P_1 P_2$	Yes

game, which specify no communication, and somehow "get together."

The equilibria of all three remaining games are included in the Pareto set. In the deadlock game, the outcome (P_1, P_2) is the second preference for each player, and both cannot improve their payoff either unilaterally (because $[P_1, P_2]$ is an equilibrium) or collectively (because the combination of payoffs is Pareto optimal).[6] In the assurance game, the reasonable equilibrium is the outcome each player most prefers. Therefore, there is no question of its Pareto optimality. Finally, in the chicken game, the only outcome that is not Pareto optimal is similar to the prisoners' dilemma (P_1, P_2); however, this outcome is not an equilibrium.

Table 3.2 summarizes the characteristics of the four games and the sources of instability of each. To summarize the argument: Nash equilibria are outcomes that are stable against unilateral deviations because no player has the incentive to deviate; Pareto optimal points are stable against universal coalitions because it is not possible to deviate from such points without hurting some players. It follows that in a two-person game, an outcome that is a Nash equilibrium and is Pareto optimal cannot be changed either unilaterally or collectively. The deadlock and assurance games have stable outcomes because both have a unique reasonable Nash equilibrium that is Pareto optimal. The unique equilibrium of the prisoners' dilemma is not Pareto optimal, and the players cannot improve their payoffs without violating the rules of the game.

6. Readers can verify these points by reversing the names of the points (P_1, P_2) and (R_1, R_2) in Figure 3.5 and transforming the prisoners' dilemma game into a deadlock.

Chicken has two equilibria, both of which are Pareto optimal, and there is no way of selecting between the two. Throughout this discussion, the equilibria of each game do not depend on the *size* of the payoffs; they depend only on the nature of each game, that is, on the *order* of payoffs.

This completes the description of the relevant game theoretic properties of the four games. Note, however, that communication between players was not permitted in this analysis. Now it is time to examine the same games when communication is permitted and players can coordinate their choice of strategies or—as it will be termed from now on—use contingent or correlated strategies.

III. Contingent Strategies and Variations of Payoffs

In this section, I argue that when contingent or correlated strategies are introduced, the properties of the four games change substantially. In particular, under certain conditions that I specify, cooperation can develop in a prisoners' dilemma game. Moreover, when correlated strategies are permitted, the likelihood of the choice of each particular strategy varies with the payoffs of the game. As I said in the introduction to this chapter, I present the argument here informally and repeat the main argument as well as additional relevant points formally and precisely in the appendices.

Let us consider the following situation: two players play the prisoners' dilemma game; when the first (the row player) chooses to cooperate, the second chooses to cooperate with probability p. Call p from now on the probability of instruction.[7] Moreover, assume that once the row player chooses to defect, the column player chooses to defect as well with probability q. From now on, q is the probability of retaliation.[8] This approach uses the concept of correlated strategies (each player uses a different strategy, according to the opponent's strategy). The approach is consistent with the weak requirements of rationality introduced in Chapter

7. I use this term because a player may think she can instruct her opponent to cooperate by providing an example.

8. I use the term to indicate that the choice of defection may provoke a reaction by the opponent.

2.[9] Indeed, this approach requires the belief in such probabilities, regardless of how well founded such a belief may be. Aumann (1974) uses a common random device to generate such correlated strategies. For example, imagine that a coin is flipped, and if it comes up heads, player 1 plays C, and player 2 plays C with probability p. If it comes up tails, player 1 plays D, and player 2 plays D with probability q.

Aumann (1974) also uses (different) subjective probabilities of occurrence of the same event as a signal to generate the coordination of strategies. Moulin (1982) makes the distinction between self-enforcing correlated strategies (that is, each player is made better off by following them) and deterring scenarios that support outcomes that are not self-enforcing and require threats to be supported. Moulin uses infinite games to generate his deterring scenarios: a player is promised that any deviation from the agreed outcome will be punished severely in the remainder of an infinite game. Generally, players can develop contingent strategies if they can communicate, if they can write down contracts, or if they can enter into repeated interaction. In each of these cases, they can use their earlier communication, their contract, or their behavior in previous rounds of the game to coordinate or correlate their strategies.

Reality is, of course, more complicated than these simplifications. For example, communication may be limited, signals may be more or less clear, promises may be made but not perfectly enforced, and opportunism may be present. Only recently have such phenomena been investigated through the study of iterated games (Bendor and Mookherjee 1987). Because of the importance of iterated games and because of their frequent use in the remainder of this book, I deal with the subject separately in Section IV. Further investigation of how contingent or correlated strategies are possible and how the probabilities p and q are generated and sustained is required. In the meantime, however, I use the concepts of contingent and correlated strategies to generate propositions for games in which such strategies are possible.

Aumann (1974, 68) proves that the use of correlated strat-

---

9. More precisely, inside subjective Bayesian game theory. See Kadane and Larkey (1982).

egies in noncooperative games can produce a wide variety of equilibria.[10] Here I use this theorem to find the necessary and sufficient conditions for the choice of cooperation by each player. In particular, utilizing the probabilities of instruction (p) and retaliation (q) makes possible the examination of all four games in the same framework.

The possibility of correlated strategies in any one of our four games leads each player to choose the strategy that maximizes his or her expected utility. Equations (3.5) and (3.6) give the expected utilities of each strategy. For reasons of simplicity, the index i is dropped from the payoffs.

$$EU(D) = T(1 - q) + Pq \qquad (3.5)$$

$$EU(C) = Rp + S(1 - p) \qquad (3.6)$$

In expected utility terms, cooperation will be chosen if:

$$EU(C) - EU(D) > 0 \qquad (3.7)$$

Rearranging terms, (3.7) is equivalent to:

$$(R - S)p + (T - P)q > (T - S) \qquad (3.8)$$

So when (3.8) holds for a player, he or she will choose to cooperate. Conversely, through (3.8), readers can make educated guesses as to when cooperation is likely. Reference to equations (3.1) to (3.4) shows that all the quantities in parentheses in (3.8) are positive for all four games. This indicates that as the probabilities p and q increase, the choice of cooperation is more likely.

Inspection of (3.8) also leads to some observations of the impact of variations of the payoffs on the likelihood of cooperation in the four games. For example, when R (the payoff for mutual cooperation) increases, the number (R − S) on the left-hand side of (3.8) increases, (3.8) is more likely to hold, and cooperation is therefore more likely to be chosen. When P (the payoff for mutual defection) increases, the left-hand side of (3.8) decreases, (3.8) is less likely to hold, and cooperation is therefore less likely to be chosen. When T

---

10. The distinction between cooperative and noncooperative games is based on the possibility of binding contracts between players: in cooperative games, such contracts are possible; in noncooperative games, they are not. Aumann (1974) proves the existence of equilibria for which the payoff vector is not contained within the convex hull of mixed strategy equilibria.

(the payoff for unilateral defection) increases, both the left- and the right-hand sides of (3.8) increase, but the left-hand side increases more slowly because T is multiplied by a number less than 1. Thus, (3.8) is less likely to hold, and cooperation becomes less likely. Finally, when S (the payoff for being fooled into cooperation by an opponent who defects) increases, both the right- and the left-hand sides of (3.8) decrease, but the left-hand side decreases more slowly because S is multiplied by a number less than 1. Thus, (3.8) is more likely to hold, and cooperation becomes more likely.

The reader is referred to Appendices A and B to this chapter for more observations and more formal proofs. Here I simply restate two of the propositions proven in the appendices.

Proposition 3.6. In a prisoners' dilemma, assurance, or chicken game, when correlated strategies are possible, the likelihood of cooperation increases with R and S (the payoffs of cooperation).

The meaning of proposition 3.6 can be better understood in reference to Table 3.1, which presents the payoff matrix of all games. As Table 3.1 indicates, R is the reward for mutual cooperation, and S is the sucker's payoff when one cooperates while the opponent defects. For the prisoners' dilemma game, both payoffs are bounded: the reward for mutual cooperation (R) cannot exceed the temptation payoff (T), nor can the sucker's payoff (S) exceed the penalty for mutual defection (P) (condition 3.1). These remarks lead us to the intuitive meaning of proposition 3.6 for a prisoners' dilemma game: cooperation is more likely when the dominance of defection over cooperation is less pronounced. For the chicken and assurance games, the reasoning is similar, though there is no dominant strategy. An increase in the rewards of cooperation makes cooperation (which is not dominated) more attractive.

The next proposition relates the likelihood of cooperation to the temptation reward (T) and the mutual penalty for defection (P).

Proposition 3.7. In a prisoners' dilemma, assurance, or chicken game, when correlated strategies are possible, the likelihood of cooperation decreases with T and P (the payoffs for defection).

Propositions 3.6 and 3.7 provide valid information for the development of cooperation *within the same setting*; across different settings, the values of p and q, the probabilities of instruction and retaliation, are too different to permit any comparisons.

It may appear that one did not need game theory to conclude that if a player's payoffs from the choice of a certain strategy increase, the player is more likely to choose this strategy. There are two reasons this objection is incorrect. First, the development of game theory has instructed us that bare intuition is an extremely important but *very* unreliable advisor, and what may seem obvious can also be plainly wrong. Second, the logic of these arguments relies on the development of correlated strategies. In order to develop such strategies, promises, threats, credible threats, or punishments are required. These concepts are important for an understanding of the behavior of political actors.[11] In the particular case of a prisoners' dilemma game, the intuition that a change in a player's payoffs will induce a change in strategy is plainly wrong if the game is single shot and no contingent strategies are permitted (as Section II indicated).

Propositions 3.6 and 3.7 hold also for the assurance and the chicken games. This remark is very important: as I said in Chapter 1 and again in the introduction to this chapter, games in multiple arenas have variable payoffs, and the variations of the payoffs in games in which contingent strategies are permitted produce the same outcomes *regardless of the nature of the game*; the outcomes depend only on the size of the payoffs.

In Chapters 5, 6, and 7, I describe situations in which actors are involved in games in several arenas and the conditions prevailing in other arenas determine the payoffs in the principal arena. Consequently, when the conditions in one arena change, the payoffs in the principal arena vary. If contingent strategies are possible, this variation will induce actors to change their strategies in the principal arena. So propositions 3.6 and 3.7 are used frequently to investigate the impact of the prevailing conditions in one arena on the strategies of political actors in the principal arena.

IV. Iterated Games

In this section, I treat iterated games separately because of their prominent position in the game theoretic literature and because they lend themselves to the development of important game

---

11. The logic of credible and incredible threats is used extensively in Chapter 6, where Labour party activists threaten to replace their representatives if they are not extreme enough.

theoretic concepts such as credible and incredible threats and promises, which I use in subsequent chapters. However, the investigation of iterated games leads to exactly the same conclusions as the study of games with contingent strategies: what matters is not the order, but the size of the different payoffs.

When players enter into repeated interaction, they are interested in maximizing their payoffs during the entire period of their interaction. Therefore, they may choose suboptimal strategies in the one-shot game if such strategies increase their payoffs over repeated play.

Imagine two players who play the prisoners' dilemma game a number of times. One can declare to the other that she will cooperate in the first round and continue cooperating as long as the other player cooperates as well; if, however, the opponent defects even once, she will defect in all subsequent interactions (Friedman 1977). If the opponent believes this threat, they both cooperate and improve their payoffs because they receive R_i instead of P_i in each interaction.

Iterations have such an important effect because they can generate correlated strategies. In an iterated game, players can choose their strategy contingent upon their opponent's choice in the previous round(s). If such a contingent choice is made, cooperation in a prisoners' dilemma game becomes a viable option.

The fact that iterated games may have different equilibria than one-shot games became known to experimental game theorists in the 1950s when mutual cooperation was the persistent outcome of repeated prisoners' dilemma games (Luce and Raiffa 1957). At the theoretical level, the existence of multiple equilibria was proven in the 1970s (Friedman 1971). However, the facts that iterated games have different equilibria than one-shot games in general and that mutual cooperation in an iterated prisoners' dilemma can be supported as an equilibrium in particular have become known to a broad audience through the work of Axelrod (1981, 1984).

The reason the discoveries that iterated games had different equilibria than single-shot games and that cooperation was possible in a finite iterated prisoners' dilemma game appeared puzzling is the so-called "backwards induction argument," which goes as follows: because the last round is known in advance, both players will defect in this last round because there is no future to influence. Given this common knowledge, both players will defect in the penultimate round. Then the decision process unravels in the same way until the first round, when both players will defect because there is no future to influence. Therefore, if the number of rounds is known, both players will play "all defect" (or ALLD, as I call the permanent defection strategy).

Axelrod (1981, 307) conjectured that if the number of iterations were not known, the results would be different. He claimed that "with an indefinite number of interactions cooperation can emerge." This conjecture is false but still widely believed to be true.

It can be shown that if the number of iterations is *finite*, even if this fact is unknown, the same argument can be made, making ALLD the appropriate strategy (Carroll 1985; Thompson and Faith 1981, 378–79). Consider playing a prisoners' dilemma game either twice or three times, but not knowing which will be the case. You can reason as follows: if I play twice, the optimal strategy is ALLD; if I play three iterations, then the optimal choice is still ALLD; therefore, no matter what the actual number of iterations turns out to be, I should still use ALLD. More generally, if both players know they are going to interact a finite number of times, there is some finite number that they both know will never be reached (say 100^{100}). If they play a one-shot game, they would both defect. In the event the game is played twice, they would both adopt ALLD (as the backwards induction argument suggests); the same strategy is adopted if the number of iterations is three or four, and so on. In fact, one can develop an inductive argument that for any finite number of rounds, ALLD is the best response of both players to each other. Therefore, no matter what the exact number of iterations turns out to be, each player will still play ALLD.[12]

The argument is general; whether or not the two players know the number of rounds, if they know that it is finite, they will play ALLD. Moreover, as Carroll (1985) remarks, because humans do not play infinite games (because "humans do not make decisions

12. It is easy to show that in formal logic, the sentence "$(p(1) = >q)$ and $(p(2) = >q)$ and $(p(3) = >q)$ and . . . and $(p(n) = >q)$" is equivalent to the sentence "$(p(1)$ or $p(2)$ or $p(3)$ or . . . or $p(n)) = >q$." The proof can be made if one thinks of $p(i)$ as the sentence "if the number of iterations is i" and of q as the sentence "I play ALLD."

about anything forever!"), the significance of the existence of cooperative equilibria in infinite iterations is problematic.

This line of reasoning eliminates cooperation altogether: in all cases, both players know that their opponent is going to use backwards induction. Therefore, they will both play ALLD. Regardless of whether the number of iterations is ten or one million, whether it is known or unknown, as long as it is finite (and this is common-sense knowledge), both players play ALLD.

Nonetheless, this is only a special case in which ALLD is the optimal choice. If your opponent plays ALLC (cooperate in every round of the game), ALLD is still preferable for you. Furthermore, if your opponent flips a coin before each round and cooperates if it comes up heads, defecting otherwise, ALLD is still optimal for you. If your opponent uses another randomizing device and cooperates 60 percent of the time, ALLD still produces the best outcome for you. Finally, if your opponent alternates C and D every five times, ALLD should still be adopted. Generally speaking, if a player knows that her opponent is going to use *any* strategy that is not dependent on her own, she will prefer to play ALLD. You should deviate from ALLD only if you believe your opponent is going to use some form of contingent strategy. This analysis precisely replicates proposition 3.5 in Appendix A of this chapter: as long as the players do not use contingent strategies, cooperation between rational, self-interested opponents in the prisoners' dilemma game cannot develop.

Essentially, backwards induction eliminates the possibility of contingent strategies because it provides each player with information about the opponent's unconditional choice of strategy: ALLD. Indeed, if you know that your opponent will play ALLD regardless of your own actions, you should prefer to play ALLD as well. Note that backwards induction is one possible reason your opponent may choose ALLD. Other reasons are also possible: she could be narrowminded or belligerent; in this case, too, ALLD is the best strategy. She could also be kind and willing to play ALLC; ALLD is the best response in this case too. So ALLD emerges as a strategy to be used not only against a rational opponent (an opponent who uses backwards induction) but also any time your opponent does not use contingent strategies.

However, all experiments indicate that sophisticated players ignore the prescriptions of backwards induction. In fact, in Axelrod's tournaments, in which political scientists, psychologists, computer scientists, and economists participated, the most successful strategies were the "nice" ones, that is, the ones that never prescribed defection first. The choice of nice strategies was common even when players knew in advance the exact number of rounds.

This discrepancy between game theoretic prescriptions and actual behavior was explained by Fudenberg and Maskin (1986), who proved a "folk theorem" about iterated games.[13] The folk theorem states, "any individually rational outcome can arise as a Nash equilibrium in infinitely repeated games with sufficiently little discounting" (Fudenberg and Maskin 1986, 533). In nontechnical terms, the theorem means that *any* outcome that gives each player no less than she could get on her own can be stable. Fudenberg and Maskin proved that the proposition is true not only for infinitely repeated games, but for games in which the number of repetitions is finite (provided this number is sufficiently large) if there is incomplete information, that is, uncertainty about the opponent's payoffs. If a player does not know her opponent's payoffs, then she can believe there is a small probability that her opponent's payoffs make mutual cooperation rational for him. In this case, she may choose to cooperate, and mutual cooperation will be the outcome of each iteration of the game. Incomplete information is sufficient to eliminate the distinction between a finite and infinite number of rounds.[14] Fudenberg and Maskin proved that a cooperative strategy could be the equilibrium outcome of an iterated game when the number of iterations is either finite (if there is incomplete information) or infinite.

The folk theorem multiplies the number of equilibria for iterated games. In fact, any individually rational outcome can be supported as a Nash equilibrium of the iterated game.[15] The con-

13. "Folk theorems" are propositions that are assumed to be true long before their formal proof.

14. See also Kreps et al. (1982), where the following argument is made: uncertainty about whether your opponent knows that you know that she knows . . . that one of the two is rational, is sufficient to generate cooperative strategies in equilibrium.

15. More precisely, as perfect equilibrium. Perfect equilibria are Nash equilibria that are not supported by incredible threats (Selten 1975).

sequences of the folk theorem for our four games are the following: in a prisoners' dilemma game, each player can guarantee herself P_i no matter what the other does. Therefore, any outcome that gives each player at least P_i can be an equilibrium. In particular, players can arrive not only at the outcome (R_1, R_2), as Axelrod's experiments demonstrated, but at any point that is Pareto superior to (P_1, P_2), that is, at any point located northeast of (P_1, P_2) in Figure 3.6. In a chicken game, each player can guarantee herself S_i. Thus, any outcome that Pareto dominates (S_1, S_2) can be the equilibrium of an iterated chicken game. Finally, in the assurance game, the individually rational (maximin) outcome for each player is P_i. Consequently, any outcome Pareto superior to (P_1, P_2) can be an equilibrium. This is an important finding because it indicates that iterated games can be used to help both players reach the Pareto surface. I use this finding in Chapter 4 when discussing the design of institutions and in Chapter 6 to explain Belgian institutions.

Iterations replaced the one or two equilibria of each game by an infinity of equilibria. Are there any possible predictions concerning iterated games, or does the infinity of equilibria make predictions meaningless? The answer to the first question is affirmative, but to explain the reasons, we must walk through Fudenberg and Maskin's proof.

Fudenberg and Maskin (1986) claim that the players can arrive at some agreement that specifies for each player a mixture of strategies. Then each player can threaten the opponent that if she deviates from the agreement, she will have to bear the maximum possible punishment. Such a statement is an actual threat only if the opponent will lose more in subsequent iterations than she stands to gain by the deviation in one iteration. So if the number of subsequent rounds is "sufficiently large," the promise of punishment is an actual threat. The proof demonstrates that there is always such a number of iterations that makes threats credible. However, the actual number depends on each player's payoffs. Consider, for example, one player in a prisoners' dilemma game against two different opponents. One of them has the payoffs $(T = 6, R = 5, P = 2, S = 1)$. Thus, if the opponent defects, he will gain 1 unit (6 instead of 5) once; from then on, he will receive only 2 instead of 5 (a net loss of 3) in each iteration of the game.

The other opponent's payoffs are (T = 6, R = 4, P = 3, S = 1). Consequently, if he defects, he will gain 2 units; from then on, he will receive 3 instead of 4 (a net loss of 1) in each iteration of the game.

Which opponent is more vulnerable to the threat of permanent retaliation (ALLD)? Obviously, the loss in the first case is much higher than in the second, and the first opponent is more likely than the second to respect the agreement. In fact, one more iteration is sufficient to make the first opponent respect the agreement, but more than two iterations are required in the second case. So when the rewards for cooperation increase or the rewards for defection decrease, smaller time horizons are required for cooperation to develop. In this sense, when the payoffs for cooperation increase or the payoffs for defection decrease, cooperation becomes more likely (cooperation requires shorter time horizons to develop).

Detailing Fudenberg and Maskin's proof would be a mathematically complicated exercise. However, the essence of their argument is no different from the arguments I made and the propositions I proved in Section III and in Appendices A and B to this chapter. The deeper reason for this coincidence is that iterated games permit the development of contingent or correlated strategies; consequently, all the propositions proven in the general case also hold in the special case.

V. Conclusions

In Section I, I explained why games in multiple arenas are games with variable payoffs and why events or moves of other players in other arenas affect the payoffs in the principal arena. In the subsequent sections, I presented four different two-by-two games (prisoners' dilemma, deadlock, chicken, and assurance), first as single-shot games without contingent strategies, then as single-shot games with contingent strategies, and finally in an iterated framework. In single-shot games without contingent strategies, the ordering of payoffs is sufficient to calculate equilibria. Of particular importance is the number of equilibria within a game because the existence of several equilibria may produce unstable outcomes. There is only one game in which the number of reasonable equilibria is greater than one: the game of chicken. Of the remaining games, the prisoners' dilemma produces a different kind of instability: the unique equilibrium is Pareto inferior. However, in a single-shot game, the players cannot improve their results.

The study of single-shot games without contingent strategies confirms Chapter 1 in the following way: it leaves the impression that rationality is a restrictive requirement. Once the requirement is imposed, the number of possible equilibria is restricted and sometimes (as in the case of prisoners' dilemma) unsatisfactory. The situation changes dramatically, however, when contingent strategies are permitted: the number of equilibria becomes infinite. In the case of contingent strategies, rationality (and game theory), far from being too restrictive, becomes inadequately so.

This important finding undercuts arguments against rationality. It is not true that actual human behavior is more various than the models assuming rationality can produce. On the contrary, the rationality assumption can produce greater variability of outcomes than exist in reality. In fact, the variability of outcomes produced by rational-choice models is so wide that one has to impose additional restrictions in order to produce a match between models and reality.

Out of the infinite number of outcomes, those in which the strategy of cooperation is chosen with higher frequency become more likely when the payoffs R and S increase, and the payoffs T and P decrease *regardless of the nature of the game*. Similar outcomes are produced in the case of iterated games. This similarity is due to the fact that iterations permit the development of contingent or correlated strategies.

The relation between the size of payoffs and the likelihood of different strategies in games with contingent strategies is given by propositions 3.6 and 3.7: the likelihood of cooperation increases when the payoffs for cooperation (R and S) increase and decreases when the payoffs for defection (T and P) increase. I use these propositions throughout this book to study games in multiple arenas, that is, games with variable payoffs in which the prevailing situation in one arena affects the payoffs of a game in the principal arena.

Appendix to Chapter 3: A

I will prove a series of propositions relating the likelihood of cooperation to the different payoffs and to the probabilities of instruction and retaliation in a prisoners' dilemma game.

The graphic representation of (3.8) in the (p, q) plane is given in Figure 3.7 and provides interesting insights into the situation.[16] In Figure 3.7, the probability of instruction p is represented along the horizontal axis, and the probability of retaliation q is represented along the vertical axis. These probabilities can take any value in the [0, 1] interval, so both can be represented by any point inside the square defined by the points 0, 1, (1, 1), and 1. I call this square the unit square. Figure 3.7 helps answer the question: "what combinations of p and q will produce the choice of cooperation in the prisoners' dilemma game?" Equation (3.8) gives the necessary and sufficient conditions for the choice of cooperation by a rational, self-interested player.

The straight line $E = 0$ represents the case of indifference between cooperation and defection. Indeed, when $E = 0$, the expected utility for each strategy is the same. $E = 0$ intersects the p axis at the point $p_1 = (T - S)/(R - S)$ and the q axis at the point $q_1 = (T - S)/(T - P)$.[17] From (3.1), it follows that each of these

16. More strictly, Figure 3.7 should represent the choice problem for both players. Therefore, a four-dimensional space would be required. Such a representation, however, is not possible geometrically, and the two-dimensional space of the figure is sufficient to produce all the conclusions required for the subsequent chapters of the book.

17. To calculate the value p_1 where the line $E = 0$ intersects with the p axis, one has to use (3.8) and set the right-hand side (rhs) equal to zero and q equal to zero. The outcome of this procedure is $p_1 = (T - S)/(R - S)$. Similarly, if one sets the rhs of (3.8) to zero and p to zero, the point $q_1 = (T - S)/(T - P)$ of the intersection of $E = 0$ with the q axis can be calculated.

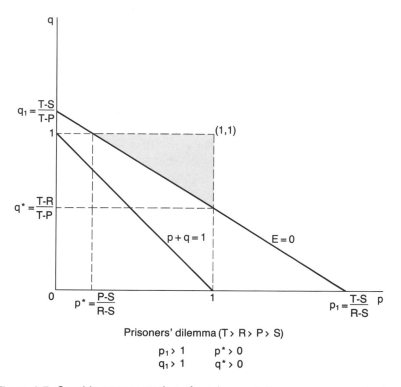

Figure 3.7 Graphic representation of a prisoners' dilemma game on the (p, q) plane.

quantities is greater than 1. The line $p + q = 1$ represents the case where the two players decide their strategies independently of each other.[18] This definition of independence is equivalent to the more familiar definition of statistical independence.[19] It follows that the line $p + q = 1$ is *always* to the southwest of $E = 0$ in Figure 3.7.

18. Indeed, along the line $p + q = 1$, p—the conditional probability of cooperation when the other cooperates—is equal to $1 - q$, that is, the conditional probability of cooperation when the other defects. So the other player cooperates or defects with the same probability regardless of what one does. In other words, each of the two players decides about her course of action independent of the other. This is precisely the case I investigated in Section I.

19. Statistical independence is defined as the case in which the conditional probability of cooperation (when the other cooperates) is equal to the unconditional probability of cooperation. However, because the unconditional probability of cooperation is the weighted average of the conditional probabilities of

To understand the practical significance of these remarks, we must return to our assumptions: that both players were rational in the sense that they sought to maximize their expected utilities and that they were self-interested in the sense that their utility functions did not include the other party's payoff as an argument. As noted, in a prisoners' dilemma game, the line $E = 0$ never intersects the independence line ($p + q = 1$). Stated formally:

Proposition 3.1. In a prisoners' dilemma game, rational, self-interested, and independent players will never cooperate.

This proposition is a simple reformulation of what we saw in Section I, that is, that defection is the dominant strategy of a prisoners' dilemma game and that the Nash equilibrium of this game is mutual defection. However, the introduction of the probabilities of instruction (p) and retaliation (q) made it possible to introduce explicitly the condition of independent players. Moreover, as will become clear in Appendix B, the introduction of the probabilities of instruction (p) and retaliation (q) makes possible a unified treatment of all four games.

Figure 3.7 can be used to calculate the minimum probabilities of instruction (p^*) and retaliation (q^*) required for the development of cooperation. The reader can verify from this figure that cooperation is possible only when $p > p^*$ (all the points in the shaded area respect this condition). Similarly, for all points in the shaded area, that is, for all points for which cooperation is possible, $q > q^*$.[20] It is easy to calculate $p^* = (P - S)/(R - S)$ and $q^* = (T - R)/(T - P)$ from Figure 3.7.[21] Both p^* and q^* are positive because of (3.1).

Proposition 3.2. In the prisoners' dilemma game, cooperation is possible only if the probability of instruction p is greater than $p^* = (P - S)/(R - S)$.

---

cooperation when the opponent cooperates and when the opponent defects, when these two conditional probabilities are equal, their average (the unconditional probability) is also equal.

20. The values of p^* and q^* are calculated as follows: set the right-hand side (rhs) of (3.8) equal to zero and $q = 1$ and calculate p^*. Set the rhs of (3.8) to zero and $p = 1$ and calculate q^*.

21. p^* can be calculated if in (3.8) we substitute 1 for q. This way we find the minimum value of p for which (3.8) holds. For the same reason, q^* can be calculated if we substitute 1 for p in (3.8).

Proposition 3.3. In the prisoners' dilemma game, cooperation is possible only if the probability of retaliation q is greater than $q^* = (T - R)/(T - P)$.

The combination of propositions 3.1 through 3.3 with inequality (3.8) gives a simple expression of the necessary and sufficient conditions for the development of cooperation in a prisoners' dilemma game.

Proposition 3.4. The necessary and sufficient conditions for cooperation in a prisoners' dilemma game are $p > (P - S)/(R - S)$ and $q > (T - S)/(T - P) - (R - S)p/(T - P)$.

All these propositions indicate a fundamental fact: cooperation will not emerge if the wrong kind of dependence between the two actors develops. If one believes his cooperative behavior will be exploited while his defection will induce his opponent to cooperate, the two players have established the wrong kind of communication for mutual cooperation.

Another implication of these propositions is the following:

Proposition 3.5. Any cooperative solution to the prisoners' dilemma game violates at least one of the three remaining conditions—rationality, self-interest, or independence of decisions.

Proposition 3.5 constitutes an algorithm for generating cooperative solutions in a prisoners' dilemma game. As I show in Chapter 4, some institutions are designed to solve coordination or prisoners' dilemma problems, and this corollary enables us to study their design.

At this point, an additional note on correlated strategies and interdependent players must be made. Traditionally, there are two branches of game theory: cooperative and noncooperative games. In cooperative game theory, players are permitted to make binding contracts; in noncooperative game theory, they are not. In terms of Figure 3.7, noncooperative game theory can be represented by the line of independence $(p + q = 1)$ because each player determines his or her course of action independent of the other. Cooperative game theory can be represented by the points $(0, 0)$ and $(1, 1)$ of these figures because binding contracts have the effect of creating conditional outcomes: all players declare they will cooperate if the other cooperates and know they will be penalized if they don't keep the promise. As the figure indicates, however, between the clear cases of cooperative and noncooperative games, there is an infinite variety of other games, in which promises can be made but kept partially, threats can be believed some of the time, or communication can fail.

I use Figure 3.7 to demonstrate the impact of variations in payoffs on the likelihood of cooperation in a prisoners' dilemma game. As noted, the shaded surface of Figure 3.7 represents all pairs of p and q that lead to cooperation in a prisoners' dilemma game. Elementary geometric calculations produce the following formula for the surface of the shaded area:[22]

$$F_{pd} = (R - P)^2/[2(T - P)(R - S)] \qquad (3.1A)$$

The impact of varying the different parameters of each game's payoff matrix on the surface of the shaded area can now be calculated. As the surface of these areas increases, more combinations of p and q (the interdependence of the two players developed in the specific setting) are sufficient to produce cooperation. This does not mean, however, that the actual values of p and q developed in a specific interaction will exceed the critical combinations specified by propositions 3.2 and 3.3. Therefore, in the best of cases, this investigation can produce average outcomes regarding the likelihood of cooperation across different settings.[23] The term *likelihood of cooperation* is used in this average-across-settings sense, indicating the surface of the shaded area in Figure 3.7.

Three additional propositions can be derived from the previous remarks.

Proposition 3.6. In a prisoners' dilemma with contingent strategies, the likelihood of cooperation increases with R and S.

The proof of this proposition can be made by checking to verify that the signs of the first derivatives of F_{pd} in (3.1A) with respect to R and S are positive.

Proposition 3.7. In a prisoners' dilemma with contingent strategies, the likelihood of cooperation decreases with T and P.

22. The shaded area is a right triangle with sides $(1 - p^*)$ and $(1 - q^*)$. Substitution of p^* and q^* from propositions 3.2 and 3.3 yields the outcome of (3.9).

23. In order to push this argument any further, one would need to make assumptions about the distribution of p and q in different settings. However, as stated in Section III, such conjectures do not exist. The implicit assumption for the remainder of this appendix is that the distribution is uniform over the unit square.

The proof of this proposition can be made by verifying that the signs of the first derivatives of F_{pd} in $(3.1A)$ with respect to T and P are negative.

All these propositions specify the conditions under which cooperation between rational, self-interested players can develop—when contingent strategies are possible. In this case, players may cooperate either because they are taught to cooperate or because they are afraid of retaliation.

Propositions 3.6 and 3.7, in conjunction with equation (3.1), indicate that for a prisoners' dilemma game, the likelihood of cooperation achieves its greatest value when the differences $(T - R)$ and $(P - S)$ tend to disappear. But these inequalities indicate that the relationship of dominance between the strategies defect and cooperate tends to disappear. Imagine, for example, a prisoners' dilemma payoff matrix with R much greater than P (from now on, $R >> P$). In this particular case, the gains from mutual cooperation may be tempting enough for players to disregard dominance. But what does $R >> P$ mean? It means that what is important for each player is the opponent's reaction, that each player is very *dependent* on the opponent. Indeed, because $R >> P$, the differences between T and R, and P and S, are minor with respect to the difference between mutual cooperation or mutual defection. What matters in this case is not so much the advantage of the dominant strategy, but whether the outcome will give a payoff of R or P.

An increase in the minimum values of p (p^*), or q (q^*) indicates that there are fewer values of p or q that satisfy equation (3.8). For example, an increase in the temptation payoff (T) indicates that only a high fear of retaliation will induce cooperation. Conversely, an increase in the sucker's payoff (S) indicates that low values of p may be sufficient for cooperation—provided the corresponding values of q are high enough to satisfy (3.8).

But when is a payoff matrix sensitive to instruction (p) and retaliation (q)? Examining Figure 3.7 indicates that if $p^* < q^*$, the interval of values of p that induce cooperation are higher than the corresponding interval of q. In other words, there will be some pairs of values p and q with $p < q$ such that the inverse couples of q and p are not sufficient for the emergence of cooperation.

Call a payoff matrix *instruction sensitive* when high values of p relative to q (that is, high p* relative to q*) are needed to induce cooperation. Call a payoff matrix *retaliation sensitive* when high values of q relative to p (high q* relative to p*) are needed to induce cooperation. What is the difference in payoffs between instruction and retaliation sensitive matrices? Conversely, given a payoff matrix, is it better to attempt instructing or threatening your opponent? By investigating the conditions of the inequality p* > q*, we obtain:

Proposition 3.8. The necessary and sufficient condition for instruction sensitive payoff matrix (p* > q*) is

$$R + P > T + S \qquad\qquad (3.2A)$$

Propositions 3.6, 3.7, and 3.8 indicate the impact of any parametric modification of the payoff matrix. In particular, propositions 3.6 and 3.7 deal with the *likelihood* of cooperation, and proposition 3.8 deals with the *reasons* for cooperation. To put it in slightly different terms, propositions 3.6 and 3.7 have behavioral implications, and proposition 3.8 has motivational content. For example, an increase of T or S is likely (proposition 3.8) to transform a prisoners' dilemma matrix from instruction to retaliation sensitive.

Appendix to Chapter 3: B

I present the graphic representation of (3.8) for the three remaining games (deadlock, chicken, and assurance) and then prove that propositions 3.6 and 3.7 hold regardless of the nature of the game.

The values of p_1, q_1, p^*, and q^* are calculated from (3.8), which holds for all four games; it follows that all games will have the same parametric expression of p_1, q_1, p^*, and q^* as functions of the payoffs (T, R, P, S). Indeed, for all games:

$$p_1 = (T - S)/(R - S) \qquad (3.1B)$$

$$q_1 = (T - S)/(T - P) \qquad (3.2B)$$

$$p^* = (P - S)/(R - S) \qquad (3.3B)$$

$$q^* = (T - R)/(T - P) \qquad (3.4B)$$

It is easy to verify that because of the defining conditions of the four games,

$p_1, q_1 > 1$ and $p^*, q^* > 0$ for a prisoners' dilemma game

$p_1, q_1 > 1$ and $p^*, q^* > 1$ for a deadlock game

$p_1 > 1, q_1 < 1, p^* < 0, q^* > 0$ for a chicken game

$p_1 < 1, q_1 > 1, p^* > 0, q^* < 0$ for an assurance game

Figure 3.8 offers a graphic representation of the deadlock game on the (p, q) plane. Because both p^* and q^* are greater than 1, the line $E = 0$ defined by (3.8) is always outside the unit square. So there are no possible values of p and q that will produce cooperation in a deadlock game.

Figure 3.9 offers a graphic representation of the chicken game on the (p, q) plane. Because p^* is negative, any value of the probability of instruction p can induce cooperation. The line $E = 0$ intersects with the independence line $(p + q = 1)$, which means

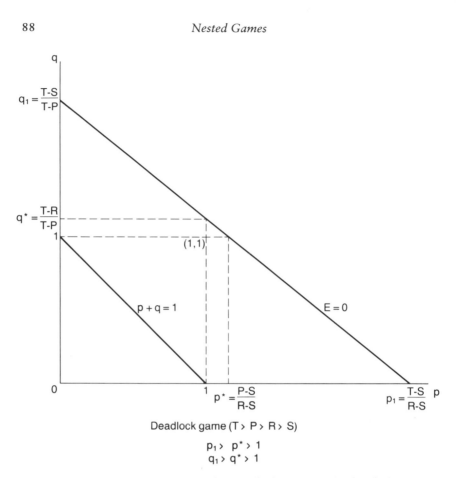

Figure 3.8 Graphic representation of a deadlock game on the (p, q) plane.

that independent players can choose to cooperate in a chicken game. The surface of the shaded area of Figure 3.9 can be calculated as the difference in the surfaces of two right triangles: the first triangle has sides $(1 - p^*)$ and $(1 - q^*)$; the second has sides $-p^*$ and $(1 - q_1)$.[24] After algebraic manipulations, the surface of the shaded area is calculated:

$$F_{ch} = (S + R - 2P)/2(T - P) \qquad (3.5B)$$

It is easy to verify that F_{ch} increases when S or R increase (because S and R appear only on the numerator of 3.5B). It is also

24. Readers are reminded that p^* is negative.

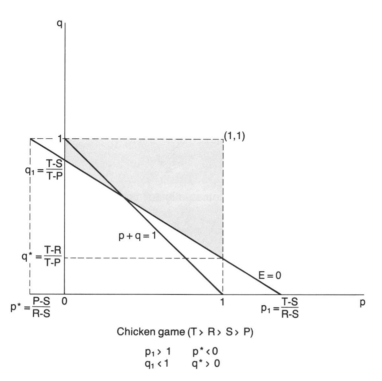

Figure 3.9 Graphic representation of a chicken game on the (p, q) plane.

easy to verify that F_{ch} decreases when T increases because T appears only on the denominator of (3.5B). It is more difficult to verify that F_{ch} decreases when P increases, because P appears with a negative sign on both the numerator and the denominator. However, the first derivative of F_{ch} with respect to P turns out to be negative (under the restrictions imposed by inequality 3.3).[25] All these variations of F_{ch} are specified by propositions 3.6 and 3.7.

Figure 3.10 offers a graphic representation of the assurance game on the (p, q) plane. Because q* is negative, any value of the probability of retaliation q can induce cooperation. The line E = 0

25. In a simpler way, because $F_{ch} < 1$ (F_{ch} is only part of the unit square), decreasing both the numerator and the denominator by the same quantity (2P) decreases the value of F_{ch}.

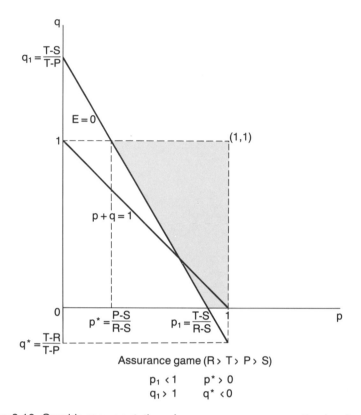

Figure 3.10 Graphic representation of an assurance game on the (p, q) plane.

intersects with the independence line $(p + q = 1)$, which means that independent players can choose to cooperate in an assurance game. The surface of the shaded area of Figure 3.10 can be calculated as the difference between the surfaces of two right triangles: the first triangle has sides $(1 - p^*)$ and $(1 - q^*)$; the second has sides $-q^*$ and $(1 - p_1)$. After algebraic manipulations, the surface of the shaded area is calculated:

$$F_{as} = (2R - P - T)/2(R - S) \qquad (3.6B)$$

It is easy to verify that F_{as} increases when P or T decrease (because T and P appear only on the numerator of 3.6B with a negative sign). It is also easy to verify that F_{as} increases when S increases because S appears only on the denominator of (3.6B) with a negative sign. A test of the first derivative of F_{as} with respect to R indicates that F_{as} increases when R increases (under the restrictions imposed by inequality [3.4]).[26] All these variations of F_{as} are specified by propositions 3.6 and 3.7.

Similarly, the derivatives of (3.6B) with respect to P and T are negative and with respect to R and S are positive. Consequently, propositions 3.6 and 3.7 hold for all three games (prisoners' dilemma, chicken, and assurance).

---

26. In a simpler way, because $F_{as} < 1$ (F_{as} is only part of the unit square), decreasing both the numerator and the denominator by the same quantity (2R) decreases the value of F_{as}.

Chapter Four

Games with Variable Rules,
or the Politics of
Institutional Change

In Chapter 2, I defined rationality as goal-oriented optimal behavior. Thus, each actor's behavior is assumed to be an optimal response to other players' behavior *and to the existing institutional structure.* Chapter 3 focused on the first kind of optimality: the mutually optimal strategies that are the subject matter of game theory. This chapter focuses on the second kind of optimality, which concerns the interaction between individuals and institutions.

The usual approach to institutions within the rational-choice tradition is to study the kinds of behavior they cause. Recurring patterns of behavior are traced back to the prevailing institutions and explained as optimal behavior under the constraints imposed by those institutions (Boudon 1984).

This chapter starts with the question of how individuals choose their optimal behavior under constraints and then theorizes about the reverse phenomenon: why and how people change the constraints of the games they play. In other words, this chapter treats institutions as endogenous and examines them as outcomes of conscious political activities.[1]

The chapter deals with cases in which one or more players try to

1. A similar approach to institutions governing labor-management relations can be found in Lange (1987).

modify the rules of a game. First, I clarify the expression *rules of the game*; as a byproduct, it will become clear that modifications of payoffs and modifications of rules are the only possible changes that can be made to a game and that consequently, games in multiple arenas and institutional change are the mutually exclusive and collectively exhaustive classes of nested games.

In game theory, a game is defined as a triplet composed of a set of players, a set of strategies for each player, and a set of payoffs for each player. The payoffs for each player are a function of the strategies each player selects. In their turn, the strategies available to each player depend on the moves available to each player, on the sequence of these moves (the order in which the players move), and on the information available before each move.[2] I call *rules* of the game the set of players, the set of permissible moves, the sequence of these moves, and the information available before each move is made. This definition is congruent with the ordinary use of the word *rule*, but it makes explicit that rules include all the characteristics of a game except the payoffs. In particular, rules include the set of players as well as the set of strategies available to each player.

According to my definition of rules, if a game varies, it is because of variations in either the payoffs or the rules (or both). Games in multiple arenas focus on the first kind of variation; institutional change deals with the second.

More explicitly, institutional change may involve one or more of the following: (1) a change in the set of players (imagine a government choosing between legislating by decree or introducing legislation on the floor of Parliament), (2) a change in permissible moves (imagine a committee introducing legislation on the floor under open rules—amendments are permitted—or closed rules—no amendments), (3) a change in the sequence of play (imagine a government asking for a vote of confidence in front of the upper chamber of Parliament before going before the lower chamber),[3]

2. Readers familiar with game theory will recognize that the triplet is a verbal translation of the definition of a game in normal form; the further analysis reflects the definition of a game in extensive form (Selten 1975).

3. In Italy, in March 1972, President Leone dissolved parliament after the defeat of the government in the senate, without waiting for a vote in the house (Allum 1973, 125).

(4) a change in available information (imagine a government declaring to Parliament that a vote on a bill would be considered a vote of confidence).[4]

Sometimes changes in rules are regulated by higher order rules. In this case, the political environment is highly structured. For example, in parliaments, strict constitutional or internal regulations prescribe which rule changes are possible and under what conditions. One can study such cases by considering the game inside the stable higher order rules. However, I present no such cases in the empirical chapters of this book.

At other times, higher order rules provide only a framework inside which actors have to move. Modification of laws inside a constitutional framework offers a prominent example. Chapter 7 provides such a case of modification of the electoral law in France.

Finally, it is possible that there is little or no explicit framework for permissible modifications; in this case, the possibilities for rule modification increase dramatically. Constitutional change is the most characteristic example. Chapter 5 deals with the change in the constitution of the British Labour party, and chapter 6 deals with constitutional changes in Belgium.

An approach to institutional change in which the rules of the game are endogenous poses several questions. Why do institutions matter? Are institutions explicitly designed, or are they the product of social evolution? Do institutions promote the interests of one actor or coalition or of the whole society?

In the following discussion, I use the term *institution* to indicate the formal rules of a recurring political or social game. The rules are assumed to be formal in order to distinguish between institutions and norms or customs. The rules are supposed to be known to the players, and each player expects every other player to conform to them. The term *recurring* in the definition is redundant because rules, even if applied only once, are always intended to cover a more or less wide range of similar cases. I choose, however, to include it explicitly because I use it frequently.

---

4. This is precisely the effect of article 49.3 of the constitution of the French Fifth Republic, as I show in Chapter 7.

Rules of the political or social game may regulate relationships among:

1. *Political actors.* Examples are the relations between government and opposition and the constitutional articles defining whether it is possible (parliamentary systems) or impossible (presidential systems) to replace the coalition in power.
2. *Institutionalized actors.* Examples include the relationships between state and federal governments or between the legislature, the executive, and the judiciary.
3. *Institutionalized actors and individual citizens.* Tax laws and the definitions of human and citizen rights that figure in prominent positions in every constitution are good examples.
4. *Individual citizens.* The right to property or the regulations included in the civil or penal codes, as well as conventions for social coordination (such as daylight savings time or driving on the right side of the road) illustrate this type of relationship.

The chapter is organized in four sections. Section I stresses the long-term character of institutions. I examine institutions as investments and indicate that time preferences are essential for understanding institutions.[5] Section II focuses on the question of the origin of institutions. Section III distinguishes two kinds of institutions: efficient institutions, which improve the welfare of all or almost all political actors, and redistributive institutions, which promote the interests of a specific coalition. This distinction is a methodological abstraction, and I show that almost all real-life institutions have a mixed character. In Section IV, I discuss the importance of distinguishing between efficient and redistributive institutions and speculate on the conditions that would make efficient institutions more likely to be selected than redistributive ones.

This chapter is less rigorous theoretically than the previous one, particularly in its discussion of redistributive institutions. This is because changing the rules of a game is an essential form of politi-

5. I borrow the concept of investments from Bates (1985). I am grateful to Robert Bates for making this very interesting paper available to me.

cal innovation and innovation defies theories and rules. I thus develop this chapter in a more inductive way and organize, discuss, and classify different ideas and theories about the creation of institutions. Each remaining chapter provides examples of different kinds of institutions and relates their character and the politics of their design to the concepts developed here.

I. Institutions as Investments

In the introduction to this chapter, I defined institutions as the formal rules of political or social games and therefore as constraints operating on individual or political actors. Each actor will try to maximize his objectives while remaining inside the institutional constraints. The problem to be solved is, therefore, one of maximization under constraints.[6] That is why I use the term *institutional design* instead of *institutional games* throughout the book. It can be shown formally that in this case, the optimal decision depends *both* on the function to be maximized (the actor's goals) and on the constraints imposed (institutions) (Theil 1968, 36–43). This obvious conclusion is the starting point for my investigation of institutions: actors maximize their goals either by changing their strategies or by changing the institutional setting that transforms their strategies into outcomes.

It has been argued that certain structures can produce certain kinds of equilibria (Shepsle 1986). With respect to legislatures, Shapley and Shubik (1954) find that it is easier to block legislation in bicameral legislatures.[7] Shepsle (1979) demonstrates that division (committees), specialization of labor (jurisdictional arrangements), or monitoring mechanisms (amendment control rules) enhance policy equilibria. Kornberg (1967) finds that limited debates or limitations of the subject matter make the "government's" program more likely to pass. DiPalma (1976) claims that direct

---

6. Game theory would have provided a more accurate description of the problem, but unfortunately, it has not been developed yet to the point where it can deal with the problem of an equilibrium when the game itself is variable, as is the case under investigation. The few attempts undertaken indicate that under limited foresight, not only the choice of strategies changes as a function of how many steps ahead a player can anticipate, but also that increasing the number of steps does not necessarily approximate the equilibrium strategies (Rice 1976).

7. See also Hedlund (1984).

power to pass legislation in committees, weak parties, no priority for government bills, and strong minorities produce legislatures unable to handle divisive and controversial issues, legislatures that pass only specific, routine legislation.

With respect to electoral laws, Duverger (1954) claimed that plurality electoral systems lead to two-party systems.[8] The issue of how votes are counted in Parliament to support either the government's bills or the government itself has tremendous importance for government effectiveness and stability: whether the constitution requires a majority of votes for or against government bills or whether Parliament can delegate legislative authority to the government have an important impact upon legislation. The presence of minority governments is made possible by rules counting abstentions in favor of the government (Strom 1984). Ballot secrecy can influence the results of votes, as is known from general elections.

The following example indicates that even small institutional details can have important and predictable results. The specific case comes from the recent political history of the Federal Republic of Germany, but many situations exist that would replicate the essential conditions. In April 1972, the government of Willy Brandt faced a motion of "constructive non-confidence" (article 67 of the Fundamental Law of the FRG). The majority supporting the government was slim, and there were fears that there would be defections in the April 27 secret ballot. This was why Mr. Brandt asked the members of his majority not to participate in the vote. This would allow him to monitor the behavior of the deputies of his coalition. The outcome was that the opposition's plans failed because they were not able to get dissenting votes from the government side and thus did not obtain the required majority (Schwartzenberg 1979).

If causal relationships, such as those provided by this example, are established between institutions and outcomes, then a political actor or a coalition of political actors may operate on the cause in order to modify its effect. For example, political actors can reduce checks and balances in order to produce more legislation, they can

---

8. For a history tracing the law's existence one century before Duverger, see Riker (1982).

modify the power of the agenda setter in order to modify the outcomes of a deliberative process (McKelvey 1979; Shepsle and
Weingast 1984), they can modify the required majorities (qualified
instead of simple) and endow certain actors with veto powers, they
can deny veto powers to certain actors and thus modify significantly the policy outcomes, and they can shift from public to
secret voting procedures or vice versa and change policies or governments.

The argument about the role of institutions is therefore pushed
one step further. I initially defined institutions as constraints.
Then I showed that because institutions systematically produce
certain kinds of outcomes, institutions can be modified to alter
policy outcomes. Knowledge of the kinds of outcomes different
institutions produce can transform preferences over policies into
preferences over institutions. Then different actors will try to select
different institutions, and in this game of institutional selection,
there will be new equilibria. To use Shepsle's words, we can pass
from institutional equilibria to equilibrium institutions.[9]

A very interesting discussion has taken place between Riker
(1980) and Shepsle (1986) as to whether preferences over institutions can come to an equilibrium or not. Both agree that majority
rule decisions about policies may lead to policy cycles, that political institutions create "structure-induced equilibria," and that
policy preferences can generate preferences over institutions. Preference over institutions, Riker argues, will generate cycles over
the choice of institutions. By contrast, Shepsle asserts essentially
that these cycles have a much longer period and therefore seem
like equilibria.

It seems to me that though both of them are correct in their
arguments, they fail to deal explicitly with a very important point:
because the expected life of institutions is much higher than the
expected life of policies, both the consequences of an institutional
choice and the uncertainty surrounding it are much more important elements in the calculation. So the transition from preferences
over policies to preferences over institutions is neither automatic
nor straightforward. For example, each citizen is obliged to file a

9. This debate about preferences over institutions can be subsumed under the
more general philosophical debate about preferences among preferences or rule
utilitarianism (Harsanyi, 1977; Jeffery 1974).

yearly tax statement. For the citizen who wants to reduce taxes over time, the following options are available: file an "honest" statement, use existing loopholes, press Congress for some additional loopholes of particular personal importance, or press for general tax reform. Each policy has expected costs, which vary with the actor's ideology, information, and probability of success—in other words, with the actor's identity. For an individual citizen who believes that the problem is personal, the choice is actually limited to the first two alternatives. For a company or a branch of industry that believes the problem concerns a broad category of people, the third may be the optimal choice. For the president of the United States or one of the two major parties, the fourth choice may be available and preferable.

However, what is more important than the difference in calculations is the difference in consequences. The first case presents an individual choice with minor consequences; the second offers a policy choice with short- and possibly medium-range consequences; the third represents an institutional change with long-term consequences.

At the theoretical level, the choice between two institutions can be considered a decision under risk that will generate two different *streams* of income because the situation is repeated over time.[10] Suppose a simple majority is sufficient to change an institution, and there is a majority coalition that would prefer a different institution over an existing one. The majority can then choose between two options. First, it is possible for actors to anticipate that an institutional change today will trigger successive modifications of institutions and to prefer to support an institution acceptable to the minority in exchange for higher institutional stability. Chapter 6 offers an example of this procedure from the Belgian constitution. Second, it is possible for the majority to decide, design, and carry out an institutional change against the will of the minority. Chapters 5 and 7 offer such cases from the political life of the United Kingdom and France.

10. The situation becomes more complicated if one considers the problem of representation of different interests by politicians and the possibilities for sincere or strategic voting of the latter in an institutionalized setting like a legislature. In the remainder of this chapter, I assume out these problems of representation and collective choice in order to focus on the creation of institutions.

The investment aspect of institutions stems from the fact that people use resources to create institutions; and once institutions are created, they generate a stream of income over time, that is, they constitute assets that can be used whenever needed in the political arena. The dilemma that follows for different players is whether they should try to make a short-term and high-return investment or a long-term one with lower rates of return. In Section III, I characterize these two different institution-building procedures as efficient and redistributive, but first I explain how we know that institution building is the outcome of conscious design.

II. Rational Choice Versus Evolutionary Explanations

"There would be no need for rules if men knew everything," claims Friedrich Hayek (1976, 21). Along the same lines, Williamson (1985) claims that institutions are required precisely because of the limited capacities of the human mind and the fact that human behavior is "intendedly rational, but only limitedly so" (Simon 1957, xxiv). In the absence of such "bounded rationality," human activity could take the form of planning because all political and social outcomes could be anticipated precisely.

There are two more conditions for the existence of institutions proposed by the economics literature: opportunism and asset specificity. *Opportunism* refers to the discrepancy between ex ante promises and ex post behavior. *Asset specificity* indicates that different actors have different endowments; therefore, they have continuing interests in the identity of one another. In the absence of opportunism, the argument goes, promises could replace institutions because one's word would be as good as one's acts. In the absence of asset specificity, market competition would replace institutions (Alchian 1984; Williamson 1985, 26–32).

Because all three conditions (bounded rationality, opportunism, and asset specificity) are frequent in real life, the study of institutions increases our understanding of social phenomena. These remarks agree with the arguments in Section I that institutions help people deal with recurring problems and situations that cannot be anticipated. This is why people *design* institutions.

My position deviates from an important body of literature that tries to explain essential aspects of human activity (either institutions or morality) in terms of evolutionary principles.[11] For example, the emergence of the state is explained as the cooperative solution to an n-person prisoners' dilemma problem (Taylor 1976). Hayek (1955, 39) claims that interesting theoretical problems arise "only in so far as regularities are observed which are not the result of anybody's design." Hayek (1979) presents such an evolutionary account in which order "emerges" out of spontaneous rules, very much the same way as order and efficiency are created in Adam Smith's market.

Let us examine these claims more closely. Suppose an agreement by the majority of the people "emerges." If this agreement promotes the interests of some people and hurts the rest, enforcement of the agreement becomes necessary, so the enforcing institution must be consciously designed. If, however, we assume symmetry among individuals, there are two possibilities: either everybody finds it preferable to respect this agreement, or everybody prefers to violate it. Choosing one side of the road to drive on or choosing time zones would be examples of the first situation; paying one's taxes would be an example of the second. In the first case, agreements are self-enforcing. In the second, they are not because each individual prefers to violate the agreement no matter what the others do. More specifically, the first situation is a pure coordination game, whereas the second can be modelled as a prisoners' dilemma game.

The only case in which an evolutionary account is satisfactory is for a pure game of coordination, that is, for a purely self-enforcing agreement. The reason is that only self-enforcing agreements do not require enforcement mechanisms.

The other kinds of institutions that can emerge, according to the evolutionary literature, are ones that solve some prisoners' dilemma problem. However, these evolutionary arguments require either some exogenous enforcement mechanism, or they deny the rationality of the agents at some step of the argument. For the prisoners' dilemma game, it has been argued that iterations can

11. For the institutional part, see Schotter (1981) or Hayek (1973, 1976, 1979); for the moral part, see Axelrod (1984) and Gauthier (1986).

make the choice of mutual cooperation possible (Axelrod 1984; Schotter 1981; Taylor 1976) and push the outcome toward the Pareto frontier. However, if in each iteration the players know the structure of the game, they also know that they will be better off if they choose to defect. In terms of proposition 3.5, *under complete information, cooperation between rational, self-interested, independent agents in a prisoners' dilemma game cannot develop.* Evolutionary arguments sacrifice the assumption of rationality. As shown in the discussion on Fudenberg and Maskin in Chapter 3, rational-choice arguments sacrifice the assumption of complete information.

For example, Axelrod (1984) argues that rationality is not required for cooperation to emerge and that even animal species or microbes "cooperate." His approach assumes perfect information. However, Axelrod's argument overlooks the fact that rationality is not only unnecessary, but is an impediment to the development of cooperation, because each rational agent knows that by deviating, she will gain.

So out of all possible "emerging" institutions, only self-enforcing agreements (pure coordination problems) can survive without enforcement mechanisms. Claims that the existence of institutions makes everybody better off are not sufficient to explain political institutions because they generally omit part of the story: enforcement of agreements. In all other cases, conscious human design and the interests lying behind it are part of the *explanandum*.

In other words, evolutionary accounts explain institutions by pointing to the interests that would lead to their creation. However, at least since Olson's (1965) seminal work, we have understood that community of interests is a necessary but not sufficient condition for people to organize and promote these interests. It is not true that "it is *only* in so far as some sort of order arises as a result of individual action but *without being designed* by any individual *that a problem is raised which demands theoretical explanation*" (Hayek 1955, 39, emphasis added). Although problems of by-products, that is, unintended consequences of human activity, need theoretical explanation in terms of the goals that were consciously promoted and led to unintended outcomes, evolutionary accounts are at best incomplete because, like other variants of functional explanation in the social sciences, they claim to "explain" an institution with reference to its beneficial consequences without the existence of a *conscious* actor. However, once a conscious actor exists, it is easy to understand why a specific institution with beneficial consequences for everyone would be created.

Chapter 3 provided a rational account for the choice of cooperation in a prisoners' dilemma game: incomplete information. In all subsequent chapters, I try to explain institutional solutions as conscious choices of the actors involved once they realize that the previously existing institutions were in constant conflict with their interests. For example, in Chapter 5, British Labour party activists repeatedly try and fail to influence party policy, until they realize that the appropriate way to have lasting influence over party policies is to redesign party institutions. In Chapter 6, in order to solve serious regional problems, Belgian elites use resources and create federal institutions that assign exclusive rights to each community to decide on issues of importance to it. In Chapter 7, after almost thirty years of stability of the electoral system in the French Fifth Republic, winning coalitions change the country's electoral system in order to remain in power.

So institutional changes may take a long time to occur, and this often creates the misleading impression of either stability or the slow evolution of institutions. However, the reason for slow institutional change is the uncertainty surrounding political institutions, which makes them similar to long-term investments, as I argued in Section I. Once political actors see that a political outcome is disadvantageous for them, they do not necessarily attempt to modify immediately existing political institutions. On the contrary, they continue working inside the same institutional framework, hoping that on the next occasion, outside conditions will work to their advantage. Only after a series of failures is a political institution likely to become disputed. Even then, however, it takes time to build political coalitions around new institutional solutions.

Although there would be neither methodological nor substantive advantage to an evolutionary account of political institutions, there would be a serious disadvantage: such an account would obscure the most important aspect of political institutions— conscious human design. In Section III, I focus on the different kinds of institutions conscious human design produces.

III. Efficient and Redistributive
Institutions

Adam Smith considers it desirable for everyone "to pursue his own
interest in his own way" (Hayek 1976, 153). The underlying reason
for this belief, aside from the well-known "invisible hand" argument, is that this constitutes a general rule that everyone can use;
therefore, no specific person or group of persons in society is systematically favored. If Marx finds this argument unacceptable, it is
not only because he does not accept the argument of the invisible
hand (in fact, his work is full of examples of self-defeating actions
inside the capitalist system), but also because he perceives a fundamental and systematic inequality, and therefore injustice, in an
exchange in which one person sells his own labor power as a com-
modity. For Smith, the exchange of labor for money, like any
other form of trade, increases the efficiency of resource allocation.
For Marx, on the contrary, the "function" of this exchange is the
creation of surplus value, the reproduction of the capitalist system
itself, and therefore, the reproduction of social inequalities.

I do not believe that the argument can be resolved at this theoretical level. The example of capitalist social relations can help us
understand that most institutions are, or at least can be examined
as, a combination of efficiency and redistribution. Therefore, I
distinguish two different ideal types of institutions, which I call
efficient and *redistributive*. The distinction is important methodo-
logically because, although different literatures examine only one
type of institution, unwarranted conclusions about all institutions
are often drawn.

I call institutions efficient if they improve (with respect to the
status quo) the condition of all (or almost all) individuals or
groups in a society. Such institutions would have the unanimous
(or nearly unanimous) support of a society. The closest example to
such an institution would be one that solved problems of coordination or of prisoners' dilemma. I call institutions redistributive
if they improve the conditions of one group in society at the expense of another. Such institutions would be supported by only
part of a society's population. The most prominent example of
such legislation is electoral laws.[12]

Figure 4.1 offers a graphic representation of efficient and redis-

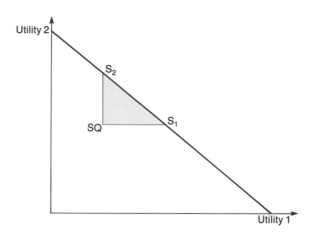

Figure 4.1 Efficient and redistributive institutions.

tributive institutions in a game with two players. If SQ represents the status quo, the shaded area represents efficient institutions because both players' utilities increase in comparison with the status quo. The rest of the diagram represents redistributive institutions because the one player's utility decreases relative to the status quo. Real political institutions are likely to fall along the borders between the two categories. Perceptions of political actors with respect to the impact of an institution vary, additionally complicating the situation. However, the most important reason that redistributive issues will arise in the design of efficient institutions is that actors will have alternative plans with respect to the desirable outcome. In Figure 4.1, for example, players 1 and 2 will prefer points S_1 and S_2, respectively; and although the choice between the previous status quo and either of these points is a matter of efficiency, the choice between S_1 and S_2 is a matter of redistribution. Despite these objections, I call any movement from the status quo inside the shaded area an efficient institution.

Arguments concerning the efficiency of institutions can be found in the works of Plato, Thomas More, and the utopian socialists, as

12. The distinction between redistributive and efficient institutions focuses on the coalition that implements the institution, not on where the initiative originates, like the "top down" and "bottom up" distinction in Banting and Simeon (1985).

well as in Adam Smith. In general, concerns with efficiency under-
lie most economic arguments. Explicit arguments about redistribution can be found in Marx, Machiavelli, Mosca, Michels, and
most authors concerned with the distribution of power in society. I
examine each ideal type separately.

1. Efficient Institutions

Why does the week have seven days? Why do drivers drive on the
right-hand side of the street? Why are there laws constraining
work on Sundays? Why does each Western European country
have one time zone? Why do we have daylight savings time? Why
did people create "the state," that is, a central agent with a
monopoly of legitimate coercive power?

The usual answer to these questions involves efficiency arguments. First, the results of a state of anarchy are compared with
the results of some kind of order. Second, it is shown that everybody becomes better off if order is imposed. Third, and here the
argument may become fast and loose (as in the evolutionary
account of institutions), order is accepted *because* it is Pareto
efficient, that is, because it improves the conditions for everyone.
Let us examine these arguments a bit more closely.

It is true that with respect to problems of coordination or prisoners' dilemma games, almost any kind of order represents an improvement upon anarchy.[13] If people are indifferent between the
possible solutions, then we will have a purely efficient institution
designed to solve problems of coordination. However, this situation is extremely unlikely.[14] More frequently, people find them-

13. I say *almost* because otherwise one could defend a brutal dictatorship on
those grounds. Indeed, although a dictatorship may represent an efficiency improvement with respect to anarchy, it may not be desired by any member of the
population. This is an important criterion distinguishing different examples of
efficient institutions. Even if anarchy is the least preferred outcome, people may
have preferences among different kinds of orders.

14. Even a seemingly innocuous problem, such as which side of the street
people should drive on, may entail disagreement (as the Swedish referendum on
this issue indicated). In fact, car owners will have a vested interest in the status
quo, and auto manufacturers will prefer a change of law. It seems to me that one
has to use extremely trivial cases (like whether to cross the street with a green or
red light) to provide examples of purely coordinating institutions.

selves facing a prisoners' dilemma type problem. I discuss the institutional solutions to prisoners' dilemma problems in detail because of their frequency and conceptual importance.

Several solutions leading to cooperation in a prisoners' dilemma game have been proposed in the literature. I summarize them all but concentrate on institutional solutions. According to proposition 3.5 in Appendix A of Chapter 3, these solutions violate the assumption of rationality, of self-interest, or of independence of players.

(1) *Violations of rationality.* Arthur Stinchcombe (1980) asks, "Why do most people do their jobs?" He answers that prisoners' dilemma games are solved "without ever having been really faced in the daily experience of most of us." His answer indicates that individuals do not try to maximize their goals, and therefore negates the rationality assumption.

(2) *Violations of self-interest.* If each player, while playing the prisoners' dilemma game, keeps in mind considerations other than self-interest, then the outcome might be mutual cooperation. In the most obvious case, each player might be concerned with his opponent's welfare. Alternatively, players may be interested in the survival of their species, as in evolutionary biology (Axelrod and Hamilton 1981; Maynard Smith 1982). In fact, evolutionary biology uses a causal force called *reproductive adaptation* that accounts for the survival of those species that maximize the number of their offspring (Elster 1983). In all these cases, it can be shown that the outcome of these altruistic considerations is the modification of the payoffs of the game. Taylor (1976) has shown that this modification may be enough to transform the game from a prisoners' dilemma to an assurance game (Elster 1978; Sen 1967).

(3) *Violations of independence.* This violation can be made in two different ways: by some kind of self-reflective or moral argument or by creation of the appropriate institutions.

(a) *Self-reflective and moral arguments.* Arguments such as Howard's (1971) metagames theory and the Kantian categorical imperative fit in the first category. Their common point is that they use some kind of mental experiment that, if accepted, solves the dilemma in favor of the cooperative solution.

Howard (1971) solves the problem by making the actors use

conditional strategies, that is, responses to the other's strategies. By repeating this intellectual experiment twice, he arrives at a situation in which mutual cooperation is the Nash equilibrium of a new game. The Kantian solution stems from the question: "what would happen if my opponent reacted the same as I?" Both solutions increase the probabilities of instruction and retaliation to their maximum value (see Chapter 3) and therefore create complete player interdependence. However, both have been criticized as irrational because, no matter what the exact argument, each player decides at the last moment independently of the other player or players. So each player's decision cannot have an effect upon the opponent. In other words, no matter what the reasoning, *it does not have any causal effect* on the opponent.

(b) *Institutional arrangements.* There are several ways that institutions can promote cooperation in a prisoners' dilemma game.

Facilitating communication and monitoring. Rapoport (1974, 18) used the example of an orderly evacuation of a burning theater to argue that there is a conflict between individual and collective rationality: collective rationality "is incorporated in every disciplined social act, for instance in an orderly evacuation of a burning theater, where acting in accordance with 'individual rationality' (trying to get out as quickly as possible) can result in a disaster for all, that is, for each 'individually rational' actor." This is a good example in which interdependence of choices can be observed (by a third person) and verified (by the actors themselves). So it is monitoring that makes the orderly evacuation possible.[15]

Permitting binding contracts. If two players confronted with a prisoners' dilemma make statements such as "I will cooperate if my opponent cooperates" and sign binding contracts to that effect, the outcome will be mutual cooperation (Myerson 1987). In real life, many sophisticated institutions have exactly this purpose: to permit people to sign binding contracts, making it impossible for them to renege on their promises.[16]

Modifying the payoff matrix of the game. The government or some other outsider could impose rewards for cooperation or

15. Barry and Hardin (1982, 385) also claim that individual and not collective rationality drives orderly evacuation.

16. Consider all the legal details and complications that permit the seller and the buyer of a house to make their transaction simultaneously.

penalties for defection. The outcome is the transformation of the prisoners' dilemma game into an assurance or a chicken game. In these cases, the noncooperative strategy is no longer dominant, and cooperative solutions may be adopted.

Transformation of the game. Shepsle and Weingast (1981) argue that the reason there are so many "pork barrel" bills in the U.S. Congress is not the practice of logrolling, but the fact that all projects are put together in omnibus bills. Indeed, logrolling in itself is less likely to produce unanimous pork barrel bills than this institutional arrangement because the latter does not permit any kind of opportunistic behavior. Each congressman retains his order of preferences; he would prefer his project be accepted and all others (equally ineffective) be voted down. But this ideal cannot be realized. He has to accept the package or kill the bill, along with his own project. In terms of the model developed in Chapter 3, $p = q = 1$. There is no possibility of defection by one without triggering the immediate and automatic retaliation by the rest. Taylor and Ward (1982) have argued that cases of lumpy public goods, like the outcomes of votes in which the decision of one individual (the pivotal voter) is enough to change the outcome, are not instances of a prisoners' dilemma game, but of chicken. Runge (1984) thinks that institutions provide better coordination because they create assurance games. So in several cases, when cooperation emerges, the game is arguably no longer a prisoners' dilemma.

Creation of an asymmetric framework. The main characteristic of such approaches is that they modify the initial game by assuming some kind of asymmetry between the two players. One of them can make a statement, or is the principal; the other responds, or adapts his behavior to the new situation. Brams (1975, 607) assumes communication between the players and distinct roles of leader and follower in order to have the "only clean escape from the dilemma."

Thompson and Faith (1981) argue that social institutions in fact provide such an asymmetric setting, in which the individuals higher in the hierarchy can make commitments, and the ones in a lower position then choose strategies. In other words, the leadership defines the rules of the game and the penalties for disobedience, and the members of the organization decide whether or not to conform to the rules. Thompson and Faith claim that most social situations (even voting) can be understood using this model. Unlike Brams, they claim that their asymmetric framework is substantially different from the prisoners' dilemma.

Iterations. As shown in Chapter 3, by reviewing the folk theorem and Fudenberg's and Maskin's (1986) proof, mutual cooperation can be a (perfect) equilibrium in iterated play of the prisoners' dilemma game when there is an infinite number of rounds or when information is incomplete.

To summarize efficient institutions: they push outcomes toward the Pareto frontier, that is, they improve the situation for all (or almost all) players. They solve problems of coordination or of prisoners' dilemma. Efficient institutions that solve coordination problems are very rare, but because agreements are self-enforcing in this case, they can be the outcome of social evolution, as shown in Section II. Noninstitutional solutions to the prisoners' dilemma have been proposed in the literature; they include violations of rationality or of self-interest and self-reflective and moral arguments. More frequent and persuasive, however, are the attempts to design institutions that eliminate prisoners' dilemma problems. Such institutions present the following characteristics: they facilitate communication and monitoring, permit binding contracts, modify the payoff matrix of the game, transform the game, create an asymmetric framework, or create an iterated play.

Chapter 6 presents a case of efficient institutions: the Belgian constitution. I show that the institutional design is quite sophisticated so as to promote mutual cooperation among elites some of the time, whereas at other times it improves social outcomes even further by leaving the power of decision making to the most concerned groups.

2. Redistributive Institutions

My treatment of redistributive institutions is less theoretical than that of efficient institutions. I do not know how far theorizing about redistributive institutions can go. The problem is that introducing one institutional change as opposed to another is a matter of political choice and coalition building. Before starting, however, we have to recognize and map the universe of redistributive institutions.

Redistributive institutions may serve two distinct purposes: either they preserve the interests of the dominant coalition, or they create a new majority composed of the previous losers and some of the previous winners. I call the first kind a *consolidating institution* because the majority is preserved and improves its position. I call the second kind a *new deal institution* because it changes policies in a significant way, and it shifts the majority. Such a shift may originate with the minority, which wants to enter into government, or it may come as an opening from one of the partners of the previously existing majority toward the minority. Figure 4.2 clarifies the distinction between consolidating and new deal institutions.

Consider a parliament with three political actors (parties), none of which has the majority of votes. What happens if an important institutional question is introduced? The introduction of an institutional issue in our society can be conceptualized as the introduction of any other (policy) issue. Figure 4.2A gives a representation of this situation in a one-dimensional issue space. Because actors 1 and 2 are closer to each other, they form the government coalition while actor 3 remains in opposition. Figures 4.2B and 4.2C are two different possible outcomes of the introduction of a new issue.

In Figure 4.2B, the introduction of the new institution is supported by the coalition in power (actors 1 and 2) and opposed by the opposition (actor 3). Actors 1 and 2 promote the new institution because it serves their interests more than the status quo (under the rationality assumption). So Figure 4.2B represents the outcome of the introduction of a consolidating institution. Chapter 7 presents such a consolidating institution in detail: the French electoral law. In Chapter 7, I return to this figure and explain how and why changes in the electoral system were made. In a general way, electoral laws are the most representative example of consolidating institutions because they have well-known redistributive properties, and they are put in place by the government, that is, the previous existing coalition.

In Figure 4.2C, the introduction of the new institution is supported by a new majority. Actors 2 and 3 find themselves closer than actors 1 and 2 after the introduction of the new institutional issue, so the previous existing coalition splits, and a new coalition is formed. The initiative for this new institution may have been

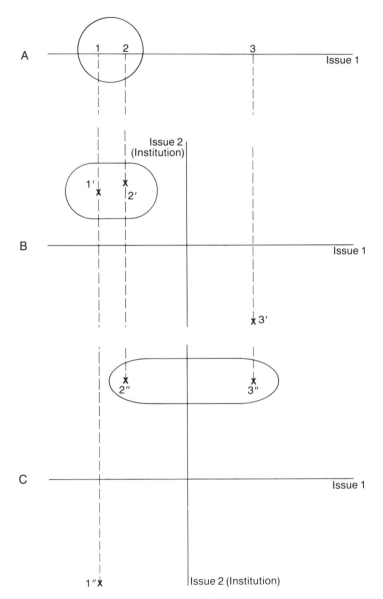

Figure 4.2A Positions of actors before the introduction of an institutional issue.
Figure 4.2B Positions of actors after the introduction of an institutional issue-consolidating institution.
Figure 4.2C Positions of actors after the introduction of an institutional issue-redistributive institution.

taken by actor 3, who was previously excluded from power *in order* to divide the previous coalition in government, or it may have been taken by actor 2, who wanted to promote his interests even further and found the alliance with actor 3 more profitable. Regardless of how the new institution was introduced, Figure 4.2C shows that it is supported by some of the previous winners (actor 2) and the previous losers (actor 3), so it is a new deal institution. Chapter 5 presents an example of a new deal institution in the internal structure of the British Labour party.

Another example of a redistributive institution of the new deal variety is the adoption of the franchise in European countries. In this case, one of the two competing factions of the dominant class, the conservatives and the liberals, were offering the franchise in order to be rewarded at the polls by the newly enfranchised (Bendix 1964; Poggi 1978; Roth 1963).

To put this discussion of different kinds of institutional design in the nested games framework, consider the game between actors 1 and 2 and the existing institutional structure (SQ). If the institutional structure changes, the two actors participate in a different game. In fact, there are several possible games that the actors can play. These games are represented by the numbers 1, . . . , n in Figure 4.3. So the actual game is nested inside a bigger game that concerns the rules of games. In a schematic representation, the game of institutional change consists of going from the institutional structure (SQ) to one of the other structures. This transition can be done in three possible ways. First, it can be done with the agreement of the actors because they find it in their common interest. In this case, the change produces an efficient institution. Second, if one of the actors has the institutional power in his hands, he can modify the institution according to his own interests. In this case, I speak about consolidating institutions. Third, the weakest actor can seek reinforcements, build coalitions, and change the institution that had not been favorable to him. In this case, I speak about new deal institutions.

Social scientists have disagreed about the nature of different institutions in general and redistributive institutions in particular. Economists, for example, explain all institutions as driven by some kind of thirst for efficiency. With respect to redistributive institutions, some social scientists believe that consolidating institutions

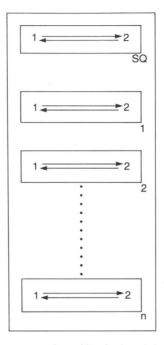

Figure 4.3 Schematic representation of institutional design.

are extremely frequent; others adopt the position that new deal institutions are essential for understanding politics. We can classify the Marxist conception of the state and institutions in the first category (consolidating institutions). Indeed, for Marxist analysis, both policies and institutions are explained by their beneficial consequences not for the whole system (as in the case of functionalism), but for the dominant class. Marx himself provides detailed analyses of why individual rights are formal and how they are in fact required for the reproduction of the capitalist system itself.[17] Institutions separating the economic and political spheres again can be explained in terms of the long-term interests of the bourgeoisie (Poggi 1978).

Riker (1983) thinks that new deal issues are the essence of politics. In his account, losers always try to bring up new issues,

17. See Marx, "On the Jewish Question," written in 1843. The text appears in Marx (1963).

disrupt existing majorities, and enter government. The possibility always exists because there is always an alternative that can defeat the status quo. In order to take advantage of this possibility, minorities have to bring up a salient issue that will create a new majority. According to his account, slavery in the United States was such a new deal issue.

In all cases, the choice of one redistributive institution over another is accompanied by public discourse concerning the public interest. In the case of the franchise, the argument can be made in an easier and more sincere way than in other cases of institutional change. In the case of changes in the electoral system in France, I show in Chapter 7 that ideological discourse was less convincing and more difficult to make. However, there are arguments in favor of different electoral systems that do not make reference to immediate instrumental partisan concerns. These include arguments for government stability (for plurality systems) and for the faithful representation of public opinion (for proportional representation).

To summarize the argument, I have claimed that institutions are of two different kinds: efficient and redistributive. Redistributive institutions can be further classified into new deal and consolidating institutions. Several discussions about institutions focus on one of these categories: economists consider only efficient institutions, Marxists consider only consolidating institutions, and liberals consider only new deal institutions. However, hasty generalizations may lead to missing the point of examining the relationships among these three categories and understanding under what conditions one kind of institution becomes more likely to occur than another. I speculate on this point instead of offering any conclusions for this chapter.

IV. Instead of Conclusion: Uncertainty and Efficiency

In this chapter, I examined three important aspects of institutions. First, institutions cannot be examined only in the short run because their implications extend over the long run. The major contribution of institutions is that they increase the stability of the political game; therefore, they facilitate the calculations of political actors.

The second point was that institutions cannot be explained through an evolutionary framework (except for the trivial case of pure coordinating institutions) because even if they did emerge out of decentralized, individual choices, they would still need centralized enforcement to be implemented.

The third point was that institutions can serve primarily the interests either of the whole society or of certain political actors. Different political (right versus left) and disciplinary (economics versus political science) traditions stress one or the other of these two kinds of institutions, which I called efficient and redistributive institutions. Most institutions represent a mixture of these two characteristics.

The question that remains to be answered is: what *kinds* of institutions are more likely to be redistributive or efficient? Consequently, when is public discourse about efficiency more likely to be well founded and persuasive?

Consider an institution with a very low expected time horizon (e.g., the electoral law in a country in which it is the custom that the majority designs its own electoral law). Compare this case with a country in which the electoral system is written into the constitution. In which of these cases is the electoral law more redistributive? The first country seems to have a clearly redistributive institution. In fact, the government is expected to design the law according to its own interests; future events can be anticipated, and the choice can be made almost safely.

In the second country, by contrast, the electoral system is expected to survive a series of elections. Can we say that its higher life expectancy indicates higher efficiency in the sense defined in this chapter? One would be tempted to answer in the affirmative. However, considering the cases of the United Kingdom and the United States, is the fact that their electoral systems remain the same an indication that they are wanted by the whole population? I think not. Even if one claimed that the half of the population of the United States that does not participate in elections prefers the current electoral system and does not abstain out of alienation, one would have a hard time making a similar argument for the United Kingdom. There, the Liberal–Social Democratic Alliance represents between one-fifth and one-third of the vote (depending on the election), and it has included the call for proportional representation in its electoral program.

Consider the laws of apartheid, or even better, the laws of slavery. Their expected and their actual time horizons were very high, but this is not a sufficient indicator of efficiency because they influenced in a systematic and clear way the balance of forces among different groups. So the discriminating factor between efficient and redistributive institutions is not how long they last (their time horizon), but the uncertainty of the outcomes they produce. Indeed, if the actors who design the institutions can foresee their consequences for different political or social groups, then they can systematically favor some of these groups. If, however, they cannot foresee the redistributive consequences, then their only guide will be the increase in the efficiency of the institution.

Finally, consider Rawls's (1971) "veil of ignorance" argument. Under the veil, individuals have to choose among societies (that is, institutions) without knowing the place they will occupy in these societies. Rawls's decision rule requires individuals to opt for the society that guarantees the most to the less privileged individual (the maximin criterion). Harsanyi's (1975) solution to the same mental experiment prescribes the choice for the society that offers the most to the average individual.[18] In general, most liberal-democratic moral arguments proceed to the adoption of rules (institutions) that are the most advantageous for a "society," which is represented as a set of undifferentiated individuals (Gauthier 1986).

It seems to me that Rawls's veil of ignorance argument is essential for the design of efficient institutions. In his case, the people who design the institution are completely uncertain about the positions they will hold in the new society, and it is their very ignorance that leads them to design a society that is better for everyone. However, it is precisely because we do not operate under this veil of ignorance, but have information (incomplete, certainly) or expectations about future events, that actual institutions are not purely efficient.

---

18. The decision-making rule is not the only difference between Rawls and Harsanyi. In addition, the first is interested in maximizing wealth; the second is interested in maximizing utilities. See Howe and Roemer (1981).

In other words, knowing precisely what kinds of outcomes an institution will produce transforms voting over outcomes into voting over institutions. Selecting institutions is the sophisticated equivalent of selecting policies or selecting outcomes. But not knowing precisely the outcomes an institution will produce makes the criteria of its ex ante adoption different from the partisan character of the outcomes it may produce.

So uncertainty or, conversely, information about the outcomes of an institution (which player one will be, in Rawls's argument) is the discriminating factor between different institutional designs. Perfect information produces redistributive institutions. Complete uncertainty produces purely efficient institutions. Both conditions are ideal types that hardly exist in reality. That is why redistributive and efficient institutions rarely exist in pure form.

The classification of institutions into efficient and redistributive completes the discussion of institutional design and the theoretical presentation of the nested games framework. It is time now to apply the theory developed in Chapters 2, 3, and 4 to political situations in Western European countries.

Chapter Five

Why Do British Labour Party Activists Commit Political Suicide?

In July 1975, the Newham North-East constituency Labour party rejected its representative, the Rt. Hon. Reginald Prentice, as official Labour candidate in the upcoming election. At that time, Prentice was a Labour cabinet minister. In the next election (1979), Prentice, representing the Conservative party in a safe Tory constituency, became a minister in the Thatcher government. His successor in Newham resigned the nomination just before the 1979 election after failing to secure adequate support for his election address from Newham Labour party activists, who felt he was insufficiently left-wing. As a result, the seat was lost for Labour in the 1979 election (McCormick 1981).

Prentice's case is unique in that it is the only occasion in British history that a constituency vote eliminated a minister. However, conflicts between constituencies and representatives are not unknown, nor was this the only case in which such conflicts led a party to defeat. For example, in 1971, Dick Taverne was disavowed by his constituency in Lincoln. Taverne resigned at once and was reelected by 58 percent of the vote, defeating both the Conservative (18 percent) and the official Labour candidate (23 percent).[1] Eddie Milne was ousted by his constituency party in

---

1. Williams (1983, 34). For a detailed and partisan account of the events, see Taverne (1974).

Blyth. Like Taverne, he ran as an Independent Labour candidate and was reelected (Bradley 1981). Margaret McKay from Clapham was forced to retire in 1970 under pressure from local activists. Her seat was marginal and was lost to Conservatives in the 1970 election (Dickson 1975). In March 1983, Bob Mellish (the former chief whip) resigned his seat in Bermondsey in response to pressure from local activists. He was replaced by the leader of these activists, Peter Tatchel, who received the Labour party nomination. In the by-election, Conservatives and former Labour supporters voted for the Liberal candidate, who won by some nine thousand votes. One of the safest Labour seats in the London area was lost (Williams 1983).

All these MPs were rejected by their Labour constituencies because they were insufficiently left-wing.[2] However, each of these replacements resulted in the loss of a Labour seat. Do Labour activists really prefer a Conservative or an Independent Labour MP (presumably hostile to Labour) over a politically moderate MP carrying the Labour banner? Do they really prefer their party to lose seats in Parliament? Why do Labour activists commit political suicide? Why do they push their party beyond the brink?

This behavior is puzzling only because I assume that political activists are rational *and* that they prefer an MP from their own party over an adversary. If one makes the opposite assumption in either of these matters, then there is no puzzle to be explained. For example, if one assumes that activists are "fanatics, cranks, and extremists," their behavior can be explained by these attributes.[3] The question then becomes why activists have these attributes.

The problem of apparent suicidal behavior does not concern British Labour activists exclusively. For example, we usually assume that political parties seek office. Indeed, the assumption is so trivial that it sometimes appears in the definition of a political party.[4] But the outcomes of events such as the 1964 or 1972 American presidential campaigns, when Barry Goldwater and

2. According to Young (1983, 2–3), similar disputes are reported in Ostrogorski. Miliband (1961, 26) finds disputes between activists and Labour party MPs as early as 1902.

3. The quotation belongs to B. Webb's diaries. See McKenzie (1964, 505).

4. See Schlesinger (1984), where a political party is defined as an "office seeking team."

George McGovern were the candidates of major parties, provoke questions about the validity of the assumptions. Similarly, the withdrawal of the French Communist party from the government and the coalition of the Left in 1984 may lead one to ask whether the Communists actually wanted to participate in government.

This chapter addresses the political dilemmas confronting the Labour leadership and activists along with the strategies chosen by each. The leadership and activists are assumed to have different political goals with at least one point in common: they both prefer Labour candidates to be elected to Parliament. Based on this assumption, the chapter attempts a rational reconstruction of the interaction between activists and leadership in the Labour party after the Second World War. On a broader scale, this chapter deals with the relationship between long-term and short-term goals and the problems of reputation and ideology, as in the case of the French Communist party. I take up this point in the conclusion.

The chapter is divided into three parts. Section I presents a simple model of candidate selection at the constituency level. The model is a game between the candidate or incumbent MP and the activists of the constituency. I show that if the game is one-shot under perfect information, readoption conflicts will not exist. Once contingent strategies become possible, or the game becomes an iterated one, however, activists' apparently suicidal behavior becomes part of a rational strategy. I then consider the iterated game as nested inside the competitive game between parties in the constituency in order to study the differences between marginal and secure seats. In Section II, I widen the picture to include the National Executive Committee and the Conference of the Labour party as additional players. Again, the game is studied as a one-shot and an iterated case, as well as simple or nested inside the competition between parties at the national level. The analysis in Section III focuses on the changes in party rules between 1979 and 1981 and the creation of the Social Democratic party (SDP). Here I argue that the major change concerning Labour was not the revision of the constitution, but the alteration in the composition of the National Executive Committee.

Throughout this chapter, I navigate a narrow strait, trying to avoid the scylla of assuming away the problem by claiming that activists are nonrational or make mistakes and the charybdis of producing a simple model that does not fit reality. I solve the problem by presenting a series of models that go from simple to more complicated; only the last model presents the full picture of the game between activists and MPs in the Labour party. However, the purpose of the intermediate models is not purely didactic. Each presents either one important part of the situation in the Labour party or an accurate description of the situation in parties of other countries. In this sense, the models are useful for comparisons between different actual or possible situations.

I. A Simple Model of Candidate Selection

I begin with two "stylized facts," that is, statements that hold not only on the average but also most of the time so that they can be considered legitimate approximations of reality. First, Labour activists are politically more extreme (that is, more to the left on the left-right spectrum) than both voters in general and voters of their party in particular. Support for this thesis can be found in Butler (1960, 5), who claims that the essential dilemma of party leaders is that "their most loyal and devoted followers tend to have more extreme views than they have themselves, and to be still further removed from the mass of those who actually provide the vote. Party leaders have to conciliate those who support them with money or with voluntary work without alienating that large body of moderate voters whose attitudes make them most likely to swing to the other party and thus to divide the next election."

Epstein (1960, 385) considers activists "zealous faithful party adherents" and states that it is the voluntary and amateur nature of constituency associations in Britain that attracts these "zealots" to the party cause, particularly at the local leadership level, where "principles not professional careers are what matter." Among scholars who have made similar remarks concerning the political opinions of party activists, one should include Ostrogorski (1892), Duverger (1952), and Key (1958). May (1969, 238) provides "a formidable array of historians, politicians, political scientists and journalists" who support the thesis of activist extremism. His list includes more than a dozen names.[5]

5. For theoretical attempts at explaining the phenomenon of activist extremism, see May (1973) and Tsebelis (1985). In disagreement with the thesis are

Additionally, more direct evidence for the thesis of constituency extremism is provided by Epstein (1960), Ranney (1965), and Dickson (1975). All have studied readoption conflicts in both major British parties, identifying several cases in which the standing MP found himself at odds with his constituency because he was too moderate and not a single case in which the standing MP was sanctioned by his constituency for being too extreme.[6] Minkin (1978, 11) claims, "For both instrumental and ideological reasons the Right was generally in favour of more independence for the PLP [Parliamentary Labour party], the Left in favour of less."

There are two possible explanations for the ideological differences between MPs and constituency activists. The first is straightforwardly electoral: MPs try to represent their constituency's median voter in order to be electorally successful. The second explanation is also essentially electoral, although the reasoning is less straightforward: it is the party position, not the personal position of an MP, that determines his reelection probability; political positions matter less for reelection than for political careers; MPs want to please the party leadership because it controls political resources; party leadership is moderate because of electoral considerations; consequently, the best way for MPs to promote their political career is through political moderation. Both arguments lead to the same outcome: moderation of MPs. For the purposes of this chapter, there is no reason to investigate the relative importance of each argument.

The second stylized fact is that constituencies are also interested in the services of their representatives. The proposition is self-

---

Rose (1962), Whiteley (1983), and Welch and Studlar (1983). Aldrich (1983a, 1983b), who has worked on Downsian models with party activists, essentially adopts the same assumption by making activists either work for their most preferred party or abstain. In order to avoid unnecessary details and criticisms, when I use the term *activists*, I refer to the General Management Council (GMC) of the constituency, which is the governing body of the constituency and is involved in the process of candidate selection. To my knowledge, although there are many complaints that this body is not representative of the constituency because it is too ideological, nobody has claimed it is not more extreme than Labour voters.

6. Dickson (1975), who covers the longest period (1948–74), finds thirty-five readoption cases in both parties, out of which eighteen were for political disagreements. Eight out of the twelve cases of Labour readoption conflict ended with the victory of the constituency.

evident. Readers can refer to Cain, Ferejohn, and Fiorina (1987) for empirical evidence. Dickson (1975) reports that seven out of thirty-five readoption conflicts in the 1948–74 period were due to the MP's neglect of constituency duties.[7]

I now simplify the selection process and formalize it at the level of the constituency. The advantage of this approach is that it explains the actors' logic in a gradual and comprehensive way. The disadvantage is that because important actors and institutions are missing from this section, the student of British politics will remain ahead of the exposition, registering objections that are not treated until Section II, where I introduce the national level in the form of additional players and study its impact on the constituency level.

In its simplest form, candidates for MP make their political statements in front of the General Management Council (the GMC or constituency activists), followed by the selection of the representative by the GMC.[8] To simplify matters further, I consider the reselection process in which the incumbent MP makes choices and sets a record for himself: this record includes his political line (moderate or extreme) and constituency service.[9] Then I consider the activists' choice between sanctions (by refusing reselection or refusing to work for him) or rewards (by reselecting and offering their personal efforts to the campaign).

With respect to constituency services, there is no problem: the more he provides, the better off he will be. With respect to political line, however, the incumbent MP faces the following dilemma: given that activists are more extreme than voters, if he is moderate, he will please the electorate even though he may lose the nomination; if he is extreme, he will get the nomination but probably will

7. Another eight were due to personal failings and the age of the MP, which can also be considered of similar nature.

8. They make a political speech lasting no more than fifteen minutes (Ranney 1965).

9. This presentation avoids issues of opportunism (e.g., the MP delivers a speech different from his record). My presentation also avoids questions of prospective and retrospective voting: do the activists believe the speech or the record? Finally, my simplification assumes out competition among different Labour candidates for selection. However, it can be defended as a reasonable approximation of the reselection process: before 1979, the incumbent MP was reselected unless the GMC objected to his record; after 1979, the rules required that a motion to continue with the standing MP be rejected by the GMC before the nomination process opened. I examine the institutional details in Section III.

lose the election. Therefore, he prefers to be moderate if he knows that the activists will not sanction him and prefers to be extreme if he knows that the activists will sanction moderation. The least preferred outcome for the candidate is to be moderate and lose the nomination.[10]

In reality, Labour MPs are aware of the trade-offs between political line and constituency service and try to use the latter in order to substitute for the former, as the following MP statements found in Cain, Ferejohn, and Fiorina (1987, 86) indicate:

— "My constituency is small enough that if you're active, the word spreads. It makes your base firm. The stronger my reputation is with the constituency association, the more I can get away with on policy."
— "Casework helps me buy some independence from the leftward drift of the party machine."
— "If the constituency party feels that the MP is neglectful, it can be very damaging. A good reputation can buy forgiveness for anything."

In each constituency and in each election, the trade-off between services and political line gravitates in one or the other direction; some constituencies will be more sensitive to political line, others to services. In this chapter, I take the trade-off for granted and focus mainly on political line. When I say that an MP was so moderate that he was replaced by his constituency, I mean the constituency was mainly concerned with political positions, the MP did not provide enough constituency service to compensate for his moderate record, or both.

For their part, the activists must keep in mind two sets of criteria when making a choice: they prefer winning the election rather than losing it but at the same time prefer their representative to be extreme rather than moderate. If the second criterion is stronger than the first, the preference order of activists is the following: (1) the most preferred outcome is the victory of an extreme representative, (2) they prefer to have an extreme repre-

10. Cases in which the candidate enters the electoral competition as an independent and wins the election despite having been rejected by his constituency party are very rare.

sentative even if he loses, (3) they prefer to have a moderate MP, (4) the worst outcome would be a moderate loser.

If this were the true preference structure of activists, they would always vote against moderate candidates, and all constituency representatives would be extreme. Because of their extremism, only a small percentage of these representatives would actually win seats in Parliament. As a result, Labour would always be the defeated party, never a party of government. A similar analysis led some Labour MPs and politicians to abandon the Labour party in 1981 and create the Social Democratic party (Bradley 1981). For those who share this analysis, the examples of lost seats in the beginning of this chapter should come as no surprise. On the contrary, these examples provide evidence supporting such an analysis. What needs to be explained is why the frequency of such incidents is so low and why Labour has not lost its place as one of the two major parties in Britain.

If activists prefer winning a seat over having an extreme representative, then their preference order is the following: (1) an extreme MP, (2) a moderate MP, (3) an extreme representative who loses, and (4) a moderate loser. This preference order assumes (as does all the existing literature) that activists are extreme, but that party interests come first.

In this chapter, I examine the consequences of both preference profiles. However, the position I defend and that explains the activists' behavior is that they have the second preference structure (they prefer a moderate MP over an extreme loser), and sometimes this preference order is known to the MP, sometimes not. In the first case, I speak about a game with complete information, in the second, about a game with incomplete information.

Figure 5.1 represents the dilemmas of candidates and activists in the form of a game. The candidate (player 1) must choose between the moderate and the extreme strategy. Subsequently, the activists (player 2) must sanction or reward. If the incumbent chooses extremism, the payoffs are R_1 for the incumbent and R_2 for the GMC. If the incumbent is moderate and the GMC sanctions, the payoffs are P_1 for the incumbent and P_2 for the GMC. If the incumbent is moderate and the GMC rewards, the payoffs are T_1 for the incumbent and S_2 for the GMC.

Let us assume a game of complete information, that is, all

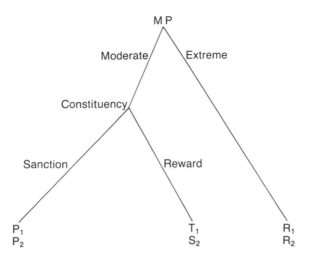

Figure 5.1 The reselection game at the constituency level.

payoffs are known to the actors, and try to duplicate the incumbent's reasoning.[11] She prefers to be moderate; yet the GMC prefers her to be extreme. Moreover, she knows that moderation may lead to the loss of selection. If moderation were going to be rewarded, she would choose to be moderate; however, if moderation were going to be sanctioned, she would choose to be extreme.

This information is available in the assumptions of our model: the first preference order of activists assumes that activists prefer to have their representative ideologically extreme rather than elected (formally, that $S_2 < P_2$); the second preference order of activists assumes that they prefer to have their MP elected rather than sanctioned (formally, that $S_2 > P_2$). Thus, if $S_2 < P_2$, the incumbent will always choose to be extreme and reselected. If $S_2 > P_2$, the MP will always be moderate, and the GMC will always acquiesce.

Note that in both cases, there is no sanction. So this model leads straight to the charybdis of empirical inaccuracy. Not only is it inaccurate, but it cannot explain the initial puzzle of the suicidal

11. More precisely, they are common knowledge, that is, they both know each other's payoffs, that each one knows that they both know each other's payoffs, that each one knows that they both know that they both know each other's payoffs. . . .

behavior of activists. In the first case, sanctioning is a credible threat and therefore is not carried out because the MP acquiesces.[12] In the second case, sanctioning is not a credible threat because it is not in the activists' interest to carry it out. However, this simple model demonstrates the logic of interaction between MPs and constituencies and explains the overwhelming majority of constituencies in the United Kingdom. After all, the apparently suicidal behavior of activists is a puzzle not because of its high frequency, but because of its mere existence.

Why are the catastrophic results in the introduction of this chapter not reproduced by this model? There are three possible answers. First, in this model, I assume that the two players' payoffs are common knowledge. If, however, I assume that the MP does not know the order of the activists' payoffs, if she misperceives their intentions, in particular, if she believes that the activists will not carry out their threat whereas in reality the GMC is unwilling to accept a moderate representative, the outcome of the reselection game is precisely moderation and sanction. This leads to the mouth of scylla twice—because it assumes the problem away by accepting that activists prefer to be defeated rather than represented by a moderate MP and because it assumes mistakes on the part of a rational actor (the MP).

The second possibility is that contingent strategies are possible or that the game is iterated; therefore, the behavior of the GMC today provides some information about its behavior in the future. In this case, it makes sense for a GMC to sanction even if $S_2 > P_2$, sending a signal to future players that it does not welcome moderation. The story of Dick Taverne represents a perfect example of signalling in an iterated game. As noted in the introduction, Taverne stood as an Independent Labour candidate against his party's official candidate and won the seat. The next election took place eight months later, and Labour won back the Lincoln seat. The activists were willing to lose the seat in one election to make the point that Taverne was unacceptable, but they were able to regain the seat with a more suitable representative eight months later. This is an acceptable answer if one assumes an infinite num-

---

12. The MP believes that if activists are asked to carry out their threat, they will because it is in their best interest to do so.

TABLE 5.1.   *The reselection game at the constituency level.*

		Constituency	
		Reward	*Sanction*
	Extreme	R_1, R_2	S_1, T_2
MP			
	Moderate	T_1, S_2	P_1, P_2

MP: $T_1 > R_1 = S_1 > P_1$
Constituency: $T2 = R2 > S2 > P2$ or
$\qquad\qquad T_2 = R_2 > P_2 > S_2$

ber of iterations of the game.[13] However, the assumption of infinite iterations is disputable.

The third possibility is a combination of uncertainty and iterations. This is the case to be investigated.[14] As noted in Chapter 3, in the case of finite iterated games with incomplete information, any individually rational outcome can be supported as a perfect equilibrium. In addition, propositions 3.6 and 3.7 indicate that the probability of which strategy is chosen varies with the magnitude of the payoffs. This information can be used to investigate the reselection game under conditions of incomplete information and iterations.

Table 5.1 replicates the two-by-two game of Table 3.1; only the names of the strategies differ. Player 1 (the MP) has the choice of an extreme or moderate record; player 2 (the GMC) has the choice

13. See the discussion of Fudenberg and Maskin's proof of the folk theorem and the backwards induction discussion in Chapter 3.

14. In the economics literature, the game in Figure 5.1 is called the chain store paradox and was first introduced by Selten (1978). The stylized story is that a chain store faces competition from a series of stores, one in each town. Each competitor must decide whether to enter the market or not, and the chain store must decide whether to apply sanctions in the form of predatory pricing or to acquiesce. The paradox occurs when, under complete information and for a finite number of rounds, the optimal behavior is to acquiesce, but an "irrational" chain store may be better off punishing its first competitor, creating a reputation that it is tough, which deters subsequent entrants. For detailed study of this and other similar games with the assumption of incomplete information, see Kreps and Wilson (1982a, 1982b) and Milgrom and Roberts (1982a, 1982b). For a study of the same game with the assumption of imperfect information, see Trockel (1986).

of sanctioning or rewarding. I have used the standard symbols introduced in Chapter 3 to represent the payoffs. In our game, the order of these payoffs is the following:

$$T_1 > R_1 = S_1 > P_1 \quad \text{for the MP} \tag{5.1}$$

$$T_2 = R_2 > S_2 > P_2 \quad \text{if the GMC makes an incredible threat} \tag{5.2}$$

$$T_2 = R_2 > P_2 > S_2 \quad \text{if the GMC makes a credible threat} \tag{5.2$'$}$$

The payoffs for the MP resemble the payoffs of a chicken game. Those for the GMC resemble either a chicken, equation (5.2), or prisoners' dilemma, equation (5.2$'$), game. The only formal difference is that one of the inequalities in each case is replaced by an equality. The reason is that, as the game in Figure 5.1 indicates, the activists do not have to respond if their MP has an extreme record. So if the MP is extreme, the outcome of the game is always the same ($R_1 = S_1$ and $R_2 = T_2$).

As noted in Chapter 3, in the iterated game, the choice of strategies depends not on the nature of the game, but on the magnitude of the payoffs. In particular, as propositions 3.6 and 3.7 indicate, when the payoffs for extremism (R_1 and S_1) increase or the payoffs for moderation (T_1 and P_1) decrease, the choice of an extreme record becomes more likely. Similarly, when the payoffs from reselection or reward (R_2 and S_2) increase or the payoffs from sanction (T_2 and P_2) decrease, rewards become more likely. We can use these findings to study the outcomes of the iterated reselection game.

Let us try to interpret propositions 3.6 and 3.7 in the context of the specific game. When is replacement of the standing MP likely, that is, when is it more probable that she will be moderate, and the activists will sanction? The answer is provided by propositions 3.6 and 3.7: the MP will be more likely to be moderate when the payoffs for extremism (R_1 and S_1) decrease or when the payoffs for moderation (T_1 and P_1) increase. The GMC will be more likely to punish when the payoffs for reward (R_2 and S_2) decrease or the payoffs for sanction (T_2 and P_2) increase. In a payoff matrix with all these characteristics, replacement of MPs by constituency activists is likely because the MP has her reasons for being moderate, but the activists signal that they will not accept a moderate representative. Let us examine the situation in different Labour constituencies to see whether these conditions are met.

To examine the payoffs in different constituencies, we have to consider the reselection game in the constituency not as isolated, but as connected to the competitive game between the two parties. So the payoffs of the reselection game vary according to the competitive situation between the two parties in the constituency. This observation is the key that not only connects the competitive situation in the constituency with the payoffs of the reselection game, but also connects the competitive situation in the constituency with the likelihood of different strategic choices by each player as well.

Consider a safe Labour seat. It is reasonable to assume that if the standing MP is moderate, she can be replaced without risk of losing the seat to the Conservatives. Thus, the value of P_2 is higher, and the likelihood of sanctioning by the GMC increases. However, the safety of the seat indicates that reselection by the GMC practically assures electoral victory; therefore, the value of P_1 decreases, leading to an increase in the likelihood of an extreme record by the standing MP. So in the case of safe seats, the probability of extreme MPs increases. This conclusion is shared by Epstein (1960, 387), who claims, "The safer the seat, other things being equal, the more vulnerable the MP to local party pressure."

Now consider a marginal seat. Similar arguments indicate that the value of R_1 decreases because the choice of an extreme record reduces the probability of winning the seat. By the same token, the value of S_2 increases because reselecting the standing (moderate) MP increases the probability of winning the seat. In this case, the choice of a moderate record and acquiescence will be more likely. Bochel and Denver (1983, 49) provide evidence of electoral considerations by activists in marginal seats.

Concerning Conservative MPs, Epstein (1960, 388) found that MPs from safe seats were removed while three MPs from marginal seats (Yates, Astor, and Kirk) survived, stressing the marginality of the constituency as a reason for their survival. In the case of Yates, Epstein (1960, 379–80) claims that he "would not have resigned if asked, and any attempt to displace him would have been politically dangerous for the conservative cause. Yates, a vigorous campaigner, had won the seat by only 478 votes in 1955, after Labour had held it in the previous three elections."

Finally, consider a safe Conservative seat. In this case, the replacement of a moderate Labour representative by an extreme one will have no impact on the probability of winning the seat. Therefore, the value of P_2 is high, the GMC is more likely to punish, and extreme representatives are more likely. Evidence for the extremism of weak constituencies is provided by Williams (1983, 28), who claims that left-wing opposition came from safe Tory seats.

Closer examination of the payoffs in different kinds of constituencies indicates that nonreselection of MPs should not be frequently observed because in marginal constituencies, the activists should be more willing to accept moderate MPs, whereas in the safe Labour or safe Tory seats, the MPs should anticipate the activists' reactions. So the actual occurrence of reselection conflict indicates two things: the existence of an iterated game with incomplete information and the existence of payoffs that make it rational in the long run for activists to send a message that they will not accept moderate MPs.

To summarize the argument, safe Conservative or safe Labour seats are more likely to have extreme representatives, and marginal seats are more likely to have moderate MPs. However, because the game is iterated, other outcomes are not excluded. In particular, it is possible, though not very likely, to observe moderation and sanctions in all three kinds of constituencies. Sanctions can be interpreted as a signal from the GMC that moderation is unsatisfactory and that in the future, prospective MPs must assume a more extreme policy stance if they want to be selected.

With this analysis in mind, what conclusions can we draw from the observed frequency of sanctions? If sanctions are used frequently, does it mean that constituencies exercise their rights more than when sanctions are not observed? Does it mean that the activists are more extreme or perhaps that candidates are more moderate? There is insufficient evidence to answer any of these questions. Earlier, I presented evidence that in the Conservative party, three MPs survived the reselection process because their seats were marginal. Similar events have not taken place in the Labour party. What inferences can we draw from this? Are Conservative constituencies driven by electoral considerations more or less than Labour constituencies? No such inference can be drawn from these data.

The reason that no inference whatsoever can be made about either player on the basis of frequency data is that they are consistent with two diametrically opposed explanations: that Labour constituencies did not bring up the issue of reselection out of electoral considerations and that Labour MPs did not get sanctioned because they anticipated that if they chose moderate positions, they would not be reselected. Because there is no way of discriminating between these two diametrically opposed explanations on the basis of the observed frequencies of sanctions, the observed frequency is meaningless.

Political scientists and journalists have struggled for a long time with the secret nature of the reselection process. Ranney (1965, 3) calls it "the secret garden of British politics." Epstein (1960, 374) says, "Studying MP-constituency association relations requires an intrusion into business which, as one MP wrote in response to my inquiry, is 'private, personal and confidential.'" However, both focus their empirical investigation on the frequencies of rejection of MPs.

This analysis demonstrates that drawing inferences from the frequency of rejection of MPs is pointless, but suggests that a significant difference is likely to characterize the political beliefs and actions of representatives from marginal and nonmarginal seats. Representatives from marginal seats will be more moderate. However, in his empirical study, Janosik (1968, 145) does not find any difference between the two. He concludes, "The nature and the extent of deviations varied little between strong and marginal constituencies." More generally, the Nuffield study of the 1979 election notes that "once again anyone looking at the record of candidate selection must be struck by how little politics entered into it" (Butler and Kavanagh 1980, 208).

On the face of this evidence, it seems that once again we sailed into the charybdis of empirical inaccuracy. My model predicts that MPs from marginal constituencies will be more moderate and will do more constituency work than MPs from safe Labour constituencies. Are these predictions false?

Without survey evidence, I could not test for the impact of marginality on constituency services. However, Cain, Ferejohn, and

Fiorina (1987, 95) did so and found that MPs from marginal seats provided more constituency services than MPs from safe seats. Using a probit model to calculate the impact of margin of victory on casework orientation, they report a highly significant negative coefficient (−.023, significant at the .01 level).

With respect to the political positions of MPs, I tested my thesis with data from the 1974–79 House of Commons. The independent variable was the marginality of the constituency: the percentage point difference between the Labour MP and the second candidate.[15] The dependent variable was the number of times each Labour MP voted with the Conservative party in the 1974–79 legislature, causing the political defeat of the Labour government. According to Norton (1980), the Labour government lost twenty-two such votes because of dissent; out of these votes, twelve were devolution bills concerning Scotland and Wales; the remaining ten constitute political defeats of the government because of dissension. The details concerning the construction of the data set are presented in the appendix to this chapter.

I correlated the frequency of dissension in nine politically sensitive votes with the margin of electoral victory of MPs in the 1974 election. The result is that the more marginal the constituency, the more likely the MP to vote with the Conservative opposition: the correlation between the margin of electoral victory and the number of times a Labour MP votes with the opposition is −.224.

Table 5.2 presents the data in summary form. Constituencies are dichotomized into "safe" (if the margin of victory is ≥20 points) and "marginal" (if the margin of victory is <20 points).[16] The MPs are divided into "loyal," if they vote with the government, and "nonloyal," if they dissent even once. Table 5.2 demonstrates that the loyal MPs come from safe seats more than from marginal seats (114 to 72); nonloyal MPs come from marginal seats more than from safe seats (83 to 63); moreover, safe seats produce more loyal than nonloyal MPs (114 to 63); and marginal seats produce more nonloyal than loyal MPs (83 to 72).

Using the same data set, I regressed the number of times a

15. The data come from Craig (1984).
16. The average margin of victory is 23.7 points; the margins are so high because the presence of third party candidates is the rule in British elections. For example, an electoral result of 50 percent, 25 percent, 25 percent produces a twenty-five-point margin of victory.

TABLE 5.2. *Frequency of dissension in the 1974–79 House of Commons, as a function of the margin of victory in the 1974 election.*

	Safe Seat (margin ≥ 20)	Marginal Seat (margin < 20)	Total
Loyal (dissent = 0)	114	72	186
Nonloyal (dissent > 0)	63	83	146
Total	177	155	332

Labour MP voted against the government (or abstained) on the margin of his electoral victory; the standardized coefficient of the variable "margin" was −.24, and it was statistically significant at the .0001 level (t − stat > 4); however, the R^2 of the model was low: .06.

The arguments can be made that statistical significance is a function of the number of cases and that, with such a high number of cases (332), statistical significance is assured, and therefore irrelevant. It is true that statistical significance depends on the number of observations and that, had I collected different data, I might not have found the same significance level. However, this argument works exactly the opposite way: the empirical test is part of the research design. Like previous studies, I could have worked with a questionnaire of a sample of MPs, which would have limited the number of cases, cast doubts on the degree their statements correspond to their attitudes or their behavior, or all three. I chose instead to examine the universe of MPs (in the 1974–79 House of Commons) and the universe of politically meaningful and instrumental dissensions (the ones that cause the Labour government to lose, so they cannot be termed symbolic).

The argument can be made that even in these votes, crucial for the Labour government, some left-wing MPs voted with the Conservative party in order to cause the government's defeat (strategic voting). Testing this argument would require data I do not have concerning left-right placement independent of cross-voting. However, in the absence of strategic voting, existing evidence strongly corroborates my theory.

However, the degree of fit of the model is low, clearly indicating that there are additional aspects of the interaction between MPs and constituencies that are not captured so far in the model. What are these aspects? In the model presented thus far, there is no central mechanism to hold the constituencies together. The description thus far fits the decentralized process of American primaries more than the British reselection process. In fact, this simple model predicts American results quite well. U.S. representatives from marginal constituencies more faithfully represent their constituencies, but representatives holding safe seats have more freedom of movement, that is, there is a "close association between electoral independence and policy independence" (Cain, Ferejohn, and Fiorina 1987, 206).[17]

II. Constituencies, the NEC, and the Annual Conference

Section I may have irritated students of British politics because it created the impression that MPs have substantial freedom of choice in their political positions, much like their American counterparts in Congress. This impression is false. The variance of opinions among British MPs is substantially lower than among their American counterparts, as the frequencies of dissension indicate. Philip Norton (1975) measured the percentage of votes in the House of Commons in which one or more of the MPs deviated from the party line and found it less than 3 percent of all votes, except for the 1959 and 1966 parliaments. In each of these cases, the government had a three-figure majority, so it could easily afford to lose some votes without suffering political or policy consequences (Norton 1975). The degree of cohesion of British parties was so high that in the 1960s, Samuel Beer (1969, 350) wrote that there was no point in measuring it. This "Prussian discipline" (in Beer's terms) withered away after 1974, as we saw in the empirical tests of Section I, but it still remained high by American standards.

The reason for the discrepancy in cohesion between Britain and the United States is the existence of party discipline in European

17. For support of the thesis, see Turner (1951), Macrae (1958), Mayhew (1966), Shannon (1968), Brady (1973) and Fiorina (1974).

countries. British MPs are rarely able to vote freely on issues of political importance. In this section, I examine one of the mechanisms that produces uniform political opinions and behavior at the national level. It requires the introduction of one additional player: the National Executive Committee (NEC) of the Labour party.

The NEC has the right to veto parliamentary candidates. Formally, this veto power is granted by clause IX, section (3) of the party constitution, which stipulates, "The selection of Labour Candidates for Parliamentary Elections shall not be regarded as completed until the name of the person selected has been placed before the meeting of the National Executive Committee, and his or her selection has been duly endorsed." More generally, the NEC has the power "to enforce the Constitution, Standing Orders, and Rules of the party and to take any action it deems necessary for such purpose, whether by way of disaffiliation of an organization, or expulsion of an individual, or otherwise."[18] Ranney describes four different forms of the NEC's veto power over GMC decisions: exclusion from list b (the list of candidates not sponsored by the trade unions), expulsion from the party, keeping names off the short list, and withholding endorsement.[19]

According to the constitution of the party, the GMC's decision must be approved by the NEC.[20] The simple model presented in Section I can now be supplemented with additional institutional details.

Figure 5.2 represents a simplified but institutionally more faithful account of the reselection game. The payoffs do not appear at the final nodes of the game tree, as is usually the case, because they would unnecessarily complicate the reasoning. The exhaustive study of the simple game at the constituency level helps us understand how this more complicated game is played by the different parties. Consider a one-shot game with perfect information like

18. Clause IX, section 2, paragraph (e) of the party's constitution (Labour Party, 1984–85, 9). Due to the continuous insertion or deletion of paragraphs, the numbers of articles, sections, and paragraphs change according to the year of publication of the constitution.

19. Ranney (1965, 154–66) finds five cases of expulsion from the party and five cases of withholding endorsement out of ten cases of NEC veto in the 1945–64 period.

20. The NEC's decision ultimately can be challenged in front of the annual conference of the party. I introduce this complication in the discussion.

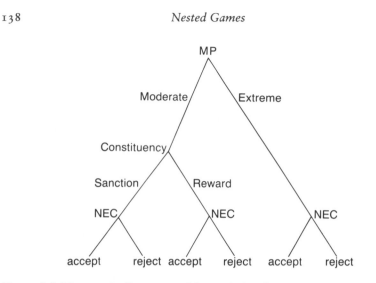

Figure 5.2 The reselection game at the party level.

the one described by Figure 5.2. Observe that the NEC, not the GMC, has the final word. Given this, in the one-shot game, the GMC is not going to make decisions that the NEC will overturn.

Two factors determine the NEC's position. The first is its political composition: if the right wing of the party controls the majority of the NEC, the GMC will recognize that it is pointless to present left-wing candidates because they will be eliminated. If the left wing of the party controls the NEC, the GMC knows the NEC will respect its choices.

The second factor that influences NEC judgment is whether the time is appropriate for the nationalization of an essentially local conflict. This second factor will temper the first one (the NEC's political composition) and will increase the range of candidates acceptable to the NEC.

These two factors jointly indicate why the empirical data presented in Section I were so noisy. The first factor indicates that the NEC is ultimately responsible for the political beliefs, attitudes, and behavior of MPs; the second indicates that the NEC will use its right to interfere according to the political situation and against extreme cases.

The incumbent can reason as follows: if the NEC is dominated by the Right, the GMC will understand that opposing her is pointless. However, even if the GMC does not understand the situation well, the NEC will put her back on the list or strike off the list all the nonacceptable left-wing candidates. In the worst case, and if the political situation is appropriate, it will threaten the constituency with disaffiliation. If the NEC is dominated by the Left, the candidate's margins are limited. If she is rejected by the GMC, she will get very little aid. For example, Prentice received a letter of support from 180 MPs but was still rejected by his constituency and the NEC (Kogan and Kogan 1982, 31).

So if the game is considered as single-shot with perfect information, there is no possibility of public disagreement between MPs, constituencies, and the NEC just as there was no possibility of disagreement between an MP and the GMC in the simplified constituency level game. The reason is that credible threats are believed and, as a result, the other player yields; incredible threats are not carried out. The game in this section is different from the game in Section I in that the NEC screen produces results that are independent of the political opinions of the constituency activists. The NEC is able through its veto power to reject any candidate it finds unacceptable on political grounds. If the NEC is dominated by the Right, constituencies acquiesce. If the NEC is dominated by the Left, MPs acquiesce. This model explains the discipline of British political parties and the overwhelming majority of reselection cases in which no disputes between constituencies, MPs, and the NEC occur. But like its counterpart in Section I, it cannot account for conflict between the actors.

If, however, the game between MPs, activists, and the NEC is considered an iterated game with incomplete information, the previous analysis (which assumed perfect information) remains basically valid, although results indicating disagreement between MPs and activists or between activists and the NEC become possible. Thus, it is possible to see constituencies supporting a candidate who is rejected by the NEC in an attempt to make a political statement for the future. Or it is possible for MPs to present views that are unacceptable to their GMC because they count on support from the NEC.

Epstein (1960) and Ranney (1965) support this analysis. Epstein (1960, 385) states:

Clearly it is the MP whose deviation from national leadership is towards the center, away from the traditional party orthodoxy, who encounters trouble in his constituency association. In addition it is doubtful that an association could punish an MP for voting contrary to local party views if, in so doing, he was following the national party leadership. Indeed when a few local Labour parties had reacted against their MPs for supporting German rearmament, the national Labour leadership intervened to protect the MPs.

But the iterated game between constituencies and the NEC is connected to the competitive game between parties at the constituency and at the national levels. If Labour is competitive in a constituency, all actors are likely to indicate moderation and try to win the seat. In this case, the payoffs for moderation increase for the MP, the constituency, and the NEC; and each actor is more likely to promote (the MP or a moderate NEC) or accept (the constituency activists or a left-wing NEC) a moderate political line and less likely to demand (the constituency activists or a left-wing NEC) or accept (the MP or a moderate NEC) extreme policies.

The number of times constituencies have yielded in front of the NEC's authority and did not present candidates who would not have survived is not known: nobody has access to the "secret garden of British politics." It is known, however, that three MPs who survived reselection conflicts with their constituencies in 1956 did so after NEC representatives threatened the constituency parties with disaffiliation (Janosik 1968, 60).

The empirical literature often addresses the question of the balance of power between constituencies and the NEC. The evidence presented is the frequency of actual vetoes: McKenzie (1964, 552) quotes Mark Abrams, who argues that "Transport House now has under its patronage hundreds of safe seats," only to disagree with him.[21] Ranney (1965) claims that the power of Transport House is more nominal than real. Nora Beloff, political correspondent, attacks Ranney in the *Observer* of August 16, 1964, writing: "A major conclusion of Ranney's inquiry is that the two parties' central organizations have much less power to place candidates

21. Journalists and politicians usually called the NEC "Transport House" because, up to the end of 1980 (when it moved to Walworth Road), it resided in Smith Square Westminster, in the same building with Britain's largest trade union, the Transport and General Workers Union.

than the public supposes. Perhaps if the professor had been tapping telephone wires linking central office, regional agents and constituency headquarters, he might have been less ready to dismiss the central office influence as 'almost nil.'" Janosik (1968, 132) attacks Beloff, claiming that for some political writers, "No matter how many charges are shown to be without foundation, those who feel that the NEC surely must have control over selection of parliamentary candidates end up insisting that the researcher did not dig deep enough and that the NEC does indeed have such powers."

My analysis suggests that the NEC has this power and that empirical evidence in the form of actual vetoes, internal memos, or phone calls is irrelevant; its power comes from the structure of the reselection game, in which the NEC has veto power over nominees. This veto power is tempered only by the general political conjuncture; if the NEC judges that the circumstances are inappropriate, it will not interfere. If this veto power is rarely exercised, it is because it rarely *needs* to be exercised. However, the veto power of the party's central leadership is always available. And the leadership can threaten to use it whenever it feels that such threats are necessary. The following infamous quote by Harold Wilson in the middle of the European Common Market crisis sums it up nicely: "All I say is watch it. Every dog is allowed one bite, but a different view is taken of a dog that goes on biting all the time. If there are doubts that the dog is biting not because of dictates of conscience but because he is considered vicious, then things happen to that dog. He may not get his licence renewed when it falls due" (*The Times*, March 3, 1967, 1).

My argument is not different from that made by the writers of the American Constitution with respect to presidential veto power. In the *Federalist*, number 74, Hamilton (Hamilton, Madison, and Jay 1961, 446) argued essentially that a veto does not have to be used in order to be effective: "A power of this nature in the executive will often have a silent and unperceived, though forcible, operation. When men, engaged in unjustifiable pursuits, are aware that obstructions may come from a quarter which they cannot control, they will often be restrained by the bare apprehension of opposition from doing what they would with eagerness rush into if no such external impediments were to be feared." So the frequency of the actual exercise of the veto cannot be used to assess relative power.

There is one more important piece of empirical evidence concerning the relative power of constituencies and the NEC. Ranney (1968, 153) demonstrates that the number of instances in which the NEC did *not* expel rebel MPs or did *not* veto left-wing candidates is much greater than the number of vetoes. Some of these cases in which veto power was politically justifiable but not exercised concern safe Conservative seats, so one can argue that the NEC did not bother to reject a politically innocuous constituency decision. Others, however, concern safe or marginal seats, where the NEC would have every reason to interfere. The most noticeable case occurred in 1955, when the Parliamentary Labour party (PLP) withdrew the whip from Aneurin Bevan and seven other MPs. In this particular case, the dispute was patched up before the NEC had to force its constituencies not to reselect them.

This empirical evidence may be interpreted as contradicting the theory developed here so far. If power lies with the NEC, why not utilize it at the most opportune time? Similarly, if the power is not used when it is needed, how do we know it exists?

These questions lead us to consider the last player in the model: the annual Labour party conference. According to the party constitution, this is the highest authority inside the Labour party; any decision can be appealed to it.

I hope that by now the reader has understood the logic of the model sufficiently to know what the impact of a fourth (and final) player in the reselection game would be. The existence of this fourth player leaves the logic of the model unchanged but may modify the results, just as the existence of the NEC modified the results but not the logic of the initial model at the constituency level.

Again, in a one-shot model with perfect information, there are no possibilities of open conflict. Even if differences do exist, each actor can anticipate the will of the majority at the annual conference: given this, the NEC knows which decisions are going to be approved. Given these two pieces of information, the GMC knows which MPs it can reject. Given all this information, MPs know what it takes to be reselected.

If reality is more complicated than this model, it is because the game is iterated, and players operate under incomplete information. In such a game, players not only play the game in each round, but are also concerned about the information they reveal to their opponent through the choice of their strategies. Though choosing a strategy that leads to an outcome that the next player will reverse may not be optimal for the current round, it sends a signal that the player wants to move to a different equilibrium and that her behavior may continue in the future. If the GMC of a safe Conservative constituency keeps rejecting moderate candidates, it may lead the NEC to accept its choice because it makes no difference for the party who the representative of the particular constituency will be. Indeed, the NEC has the choice between acquiescing and expelling the whole constituency, and in some circumstances, the second choice may be costly.

Finally, this iterated game is connected to the competitive game between the two parties at both the constituency and the national levels. Therefore, variations in the parties' competitive situation modify the payoffs of the game, influencing the likelihood that each actor will veto the decision of the previous one. In marginal constituencies, politics is more likely to be moderate because in such constituencies, the payoffs of the moderate players will increase, and the payoffs of the extreme players will decrease, making the selection of moderate strategies and acquiescence more likely.

If the NEC vetoed and severely punished MPs in the late 1940s, it is because the party was at the peak of its strength after the war and could afford to lose seats. Moreover, rebel MPs were sympathetic to the Communist bloc. The NEC would have been justified in not taking action against Bevan and his cohorts because the party was weak in 1955 and could not afford to lose MPs. Moreover, Bevan and his friends were strong inside the party, and their expulsion would have created serious problems in the annual conference.

Like the model in Section I, empirical tests of this more complicated model would require data concerning marginality, political favors, and political positions. In addition, information concerning different cohorts of MPs who were elected under different conditions of competition between Labour and Conservatives and different compositions of the NEC, and the number of times they have been reelected, would be required. I do not test the more complicated model here.

One case that indicates how changes in the situation (that is, in the payoffs of the players) can influence the final outcome is reported by Rush (1969). John Palmer was selected to represent Croydon North-West in 1962 and was accepted by the NEC. In 1966, he was reselected by his constituency, but the NEC refused to endorse him. The constituency protested at the 1966 annual conference, but was defeated. The NEC "imposed its own candidate" (*The Guardian*, March 10, 1966). Rush (1969, 140) speculates, "Whether this was an oversight, a reflection of the increasing anxiety that the serious illness of Hugh Gaitskell was causing the Labour party, or whether Palmer's left wing views were more in evidence during and after the 1964 election, it is not clear." All of Rush's explanations (with the exception of "oversight," because there is no such entry in a rational-choice dictionary) reflect some modification of the payoffs of the NEC.

To recapitulate and conclude, there are four relevant players in the reselection game. If the game is one-shot with perfect information, there is no possibility of open disagreement among actors. This does not mean that there is agreement as to which solution is optimal. It simply means that the actor with the last move can impose her will on the rest. If the game is iterated with incomplete information, open disagreements become signals from the actors indicating their desire to move to a different equilibrium. In this iterated game, the value of different payoffs influences the likelihood of players adopting different strategies. Because the reselection game is nested inside a competitive game between parties, the value of these payoffs depends on the competitive situation between parties at the constituency and national levels. In general, MPs from marginal constituencies are more moderate and more service oriented.

With respect to the internal balance of the party, one can make the following remarks: if the party is strong at the national level, the internal balance tips toward the annual conference and the NEC. If the NEC has the same political majority as the annual conference, the NEC becomes the final player, increasing its power tremendously. If the GMC is congruent with the NEC and with the annual conference, the GMC has the last word. So MPs who move to the right of the party line risk finding themselves opposed not only to the activists of their constituency, but to the NEC as well. In this case, the NEC will not try to save them from the wrath of their GMC. This explains why MPs who defied the party line from moderate positions were refused readoption by the activists of their constituency without complaint from the NEC. These are some of the contextual reasons that shift the balance of power from one actor to the other. Institutional factors, however, influence the final outcomes as well. This is the focus of Section III.

III. Balance of Power and Institutions

The political differences between the parliamentary group as a whole and constituencies become more visible during the annual conference of the Labour party.[22] In the 1950s and 1960s, the annual conference voted resolutions supporting the leadership of the party (that is, the leader and the parliamentary group) because the majority of the conference was dominated by the vote of the trade unions (which were largely moderate and represented more than 80 percent of the vote). The major exception to this rule was the Scarborough conference of 1960, when Hugh Gaitskell was defeated on the issue of unilateral disarmament. He used all his power to reverse the decision in 1961.

In the 1970s, however, the situation slowly started to change: the trade unions shifted to the left. According to Williams (1983, 30), there were three major reasons for this modification. First, there were changes at the top of the two larger unions. The turn was accidental in the case of Transport and General Workers, marginal but sufficient to alter the leadership in the Engineering Workers. Second, traditional manual workers shifted to the left in reaction to the contraction of their industries. Third, and most important, the balance inside the trade union bloc moved toward public sector and white-collar unions. The active minority in these unions was much more ideological than were workers in traditional unions.

These changes were reflected both in the resolutions of the

22. For an excellent description of the procedures inside the annual conference, see Minkin (1978).

annual Labour party conference and in the composition of the NEC. The NEC is composed of twenty-nine members, of whom seven represent the constituencies and twelve the trade unions. In 1964, only the constituency representatives were left-wing; the balance of forces was eight for the Left and twenty for the Right.[23] By 1974, the majority of the trade unions had joined forces with the constituencies, reversing the balance of forces so that fifteen were now for the Left and fourteen for the Right. This domination by the Left was confirmed in 1978: eighteen against eleven (Finer 1981, 114).

Party leadership, however, pursued moderate policies both in government and in opposition. In the late 1960s, Harold Wilson applied a strict incomes policy with declining success while in government. But the document "In the Place of Strife" created the most important policy clash between the leadership and members. The employment secretary, Barbara Castle, composed a document proposing the structural and operational reform of the trade unions. The government attempted to legislate the proposals despite the objections of the NEC, the trade unions, and their rank and file. The government did not stop until it was certain that the votes in Parliament were lacking (McLean 1978). Butler and Pinto-Duschinsky (1971, 267) argue that these incidents are related to a decline in party membership during the late 1960s.

Furthermore, party leadership chose to ignore the resolutions of the annual conference. Constituencies introduced resolution after resolution trying to induce the PLP to conform to conference decisions.[24] In 1973, Labour's program proposed the nationalization of twenty-five additional industries. Soon after the document's publication, Harold Wilson made it clear that he had no intention of applying this policy if elected. Similarly, while the program was becoming more and more radical, James Callaghan's 1979 cam-

23. At the time, the NEC had twenty-eight members.
24. For example, in 1970, one resolution was introduced asking the government to "avoid the fateful error of ignoring the wishes and voices of the constituencies and affiliated organizations from whom springs the support and strength of the Labour party movement" (Labour Party 1970, 171). Another claimed, "This conference believes that the Parliamentary Labour Party leaders, whether in government or in opposition, should reflect the views and aspirations of the Labour and trade union movement, by framing their policies on annual conference decisions" (Conference Report 1970, 180).

paign was unabashedly moderate. The most accurate description of Labour's campaign was presented in a *Sunday Mirror* cartoon, in which a caricature of Mr. Callaghan said, "If you must have a conservative P.M., I'm your man."

The parliamentary group was able to ignore the party members and the annual conference because it has a double source of legitimacy and power. Besides representing the membership, the parliamentary group represents the voters, that is, the electorate in general.[25]

For the average activist, however, the situation developed as follows: when Labour was in power, the parliamentary group made decisions that blatantly conflicted with the rank and file's will, citing economic constraints as justification. When in opposition, it made statements equally unacceptable in order to please public opinion. Activists began to realize that the supposedly binding conference resolution was full of sound and fury and signified nothing. Minkin (1978, 330) reports that for Labour party members, the government "revoked its mandate commitments, ignored conference decisions, and carried through policies which ran counter to some of the basic principles of the party." In 1973, after Wilson's statement of his intent to ignore the conference's resolution on nationalizations, the most influential group of the "outside left" was formed: the "Campaign for Labour Party Democracy" (CLPD).

The outside left differed from the inside left (groups of MPs like the "Tribune group") in that its goal was to persuade and mobilize the party's rank and file. Indeed, the outside left's success in the constituencies and the trade unions was spectacular: in 1974, the CLPD had 4 constituencies and no trade union branches as members. By 1977, the number of constituency branches had reached 74, and the number of trade union branches had grown to 25. By 1980, the numbers were 107 and 112, respectively (Kogan and Kogan 1982, 42).

The CLPD strategy was simple: instead of deserting the party (as activists did in the late 1960s) or writing slogans and manifestoes (an early 1970s strategy), left-wing activists spent their time

---

25. McKenzie (1982) claims that representation of the electorate has priority over representation of the constituency.

and energy drafting constitutional amendments. Their reasoning was that if policy is formulated de facto by the parliamentary group and the leader, one must control the parliamentary group and the leader in order to make fundamental policy decisions.

There were two main targets of the outside left. To control the parliamentary group, it wanted to transform the reselection process in order to increase MP dependence on the GMC. To control the leader, it wanted to have him elected not by the parliamentary group, but by a different electoral college in which the rank and file's opinion would be taken into account. The following discussion focuses on the first target, the "mandatory reselection" achieved in the Brighton conference in 1979.[26]

The CLPD used "model resolutions" to coordinate constituency support on particular key issues. Because each constituency has the right to send one resolution to the party conference, the CLPD used its newsletter to publish the text that sympathetic constituencies could submit to the conference. Mandatory reselection was submitted twelve times in 1975, forty-five times in 1976, seventy-nine times in 1977, sixty-seven times in 1978, and twenty-two times in 1979 (by that time, other issues were also important) (Kogan and Kogan 1982, 28).

The mandatory reselection proposition was adopted by the Brighton Conference (1979) after Labour's electoral defeat. According to Ron Hayward, the party's general secretary: "The reason [for the electoral defeat] was that, for good or ill, the Cabinet supported by MPs ignored Congress and Conference decisions. It was as simple as that . . ." (Butler and Kavanagh 1984).

The mandatory reselection as it was voted by the Brighton conference consisted of a special convention of the General Management Committee between the eighteenth and the thirty-sixth month following the election "to consider whether or not the Member of Parliament shall be selected as the prospective Parliamentary candidate" (clause XIV, section 7, paragraph [a]).

---

26. The second target, the election of the leader by an electoral college composed of 30 percent constituencies, 30 percent MPs, and 40 percent trade unions, failed in 1979. A third goal was achieved in 1979—the party's program was to be drafted by the NEC.

(e). The special meeting referred to in paragraph (a) above must consider a resolution to appoint the Member of Parliament as the prospective candidate and if the resolution is carried, the name of the Member of Parliament shall be submitted to the National Executive Committee for endorsement. . . .

(f). If the resolution referred to in paragraph (e) above is not carried the said meeting must consider a further resolution that this party shall set in motion the procedure for the selection of a prospective Parliamentary candidate in accordance with section (3) of this clause.[27]

Clause XIV, section 7 makes the reselection of the MP the rule rather than the exception, as it had been before. Before, the GMC had to mobilize against its MP; now the MP requires the support of the GMC. Before, the GMC needed to provide the NEC with reasons for not readopting its MP. No such reasons are required now. Therefore, this formula dramatically reduces the costs of rejecting the MP.

It has been observed in the past that similar rule modifications may have had spectacular impact, especially when the masses are involved. Duverger (1954) found that between 1927 and 1946, the number of members of the Labour party declined dramatically. This was due in large part to the fact that the existing "contracting out" rule for membership in parties was replaced by the "contracting in" rule, requiring workers to sign a form in order to pay their contributions and become members of the Labour party, rather than signing a form if they wished not to join. Membership surged again with the readoption of the contracting out system in 1946. Similarly, there was a ten percentage point decrease in electoral participation in the Netherlands after 1967, when voting ceased to be compulsory (Crewe 1981, 241). One might expect that in the case under discussion, lowering the costs of reselection might lead to higher mass mobilization and to different majorities and outcomes altogether.

This, however, was not the intention of the authors of the resolution, nor was this the outcome. The selection process remains in the hands of the General Management Committee, which is a

27. Report of the Seventy-eighth Annual Conference of the Labour Party, Brighton (Labour Party 1979, 444).

restricted (and some claim extremely biased to the left) subset of the constituency. The clause was not introduced to increase mass participation, and for the restricted group of people who participate in the reselection process, the costs of participation were very low to begin with. So the institutional change does not imply a compositional change of the electoral college in the constituency. Although it does make rejection easier, there is no reason to believe that the members of the GMC had been intimidated before by the difficulty of the process.

Moreover, the change does not imply a change in parliamentary personnel. In fact, refusal of reselection should be the exception rather than the rule because, as I showed in Section I, it indicates that the MP did not recognize the existing balance of forces. As Chris Mullin of the CLPD put it: "The purpose of reselection was to change the attitude of MPs, not necessarily to change MPs. There has already been a noticeable change in their attitudes" (Williams 1983, 43). Williams also provides evidence that MPs who had already been reselected were voting differently from their colleagues who were coming up for reselection. Given this analysis, he concludes, "Potentially, therefore, the CLPs could affect the votes of 60 percent rather than 30 percent of the college" for the selection of party leader (Williams 1983, 45).

There has been a noticeable change in the policies and (after the schism of the SDP) the composition of the Labour party. However, the causal process indicated by the previous quotations is incorrect. The change in policies is not the result of the change of reselection procedures and the more prominent role of activists in them. We can use the model developed in this chapter to examine closely the causes of the political shift to the left and the change in the parliamentary group of the Labour party. Examining Figure 5.2, readers can verify that clause XIV, section 7 modifies the procedure in the middle of the game tree, that is, it modifies the payoffs of the game, but does not modify the game tree itself. Indeed, the changes will make rejection of moderate MPs easier for the GMC. Moreover, it may avoid a serious confrontation between MP and constituency activists that could discourage Labour voters. However, these rules do not change the subsequent process: the GMC's decision is still subject to NEC approval. Only to the extent that the NEC, for its own political reasons, decides not to interfere in local politics do the constituencies have relative freedom of movement.

Given that the game remains essentially the same, the reader should not expect any modification of outcomes if the game is single-shot with perfect information. All outcomes come with the approval of the NEC and the annual conference. If, however, the game is considered iterated with incomplete information, the modification of payoffs introduced by clause XIV, section 7 will have some impact in shifting the balance of forces in favor of the GMC. Even in an iterated game, however, the impact of this change is marginal compared to a change in the majority within the NEC and the annual conference.[28] Indeed, changing the composition of the last two players of the reselection game changes the outcomes of the game. Because the MPs have to be approved by the NEC, as a general rule, a right-wing NEC will accept moderates and reject radicals, and a left-wing NEC will adopt the opposite policy.[29] Thus, the important change characterizing the Labour party (at least with respect to the selection of MPs) is not the change of institutions, but the change of the majority within the NEC and the annual conference. To use Paterson's (1967, 41) terminology, it was no longer the case that "the right-wing heart of the party beats loud and strong."

For my account to be complete, I need to answer two questions. (1) If the essential aspect is the change in the balance of forces in the NEC and not the institutional change, why did the CLPD fight so hard for mandatory reselection? Why did they fight for institutional change when they had already won the essential battle, that is, when they had already captured the NEC and the annual conference? (2) Why did they not introduce a more effective institutional change if the balance of forces was so favorable?

The answer to the first question is straightforward: the CLPD fought for these institutional changes because they have an impact

28. Criddle (1984, 220–25) counted eight MPs who were deselected before the 1983 election. Butler and Kavanagh (1984, 53) report that "*only* eight MPs actually lost their seats *owing* to mandatory reselection" (emphasis added). The reader by now knows that the actual number of deselected MPs is not a relevant indicator for understanding the importance of the process and that mandatory reselection was not essential for these rejections of MPs by constituencies.

29. This general rule, however, is subject to the conditions prevailing at the national level.

(small but noticeable) on the balance of forces between the Left and the Right inside the party because clause XIV, section 7 shifts the balance of forces in favor of the GMCs and appears to be more stable than temporary majorities. Moreover, depending on the prevailing political conditions, the NEC may judge that intervention at the local level may have an overall negative impact for the party and may not exercise its veto power.

The answer to the second question is more complicated: the CLPD did not introduce more effective changes because changes were not feasible at the time (if ever) for the Labour party. The most radical change would have been to abrogate the NEC veto power altogether. Such a change would elevate the constituency to the role of ultimate judge of MP behavior (see the analysis in Section I and the game in Figure 5.1). Such a modification would have meant the Americanization of the Labour party in that there would have been no central authority in the party. The GMC of each constituency would have been able to select the candidates of their choice without confronting any obstacle, and in the long run, the parliamentary group would have been composed of MPs without any common links but who faithfully reflected the GMC of their constituency.

Instead of fighting the battle around decentralization (constituencies versus central leadership), the CLPD chose to fight around the question of who is in control at the central level: the conference and the NEC or the parliamentary group. This choice was strategic: the CLPD could not count on trade union support on the decentralization question. The trade unions dispose block votes and control the majority of the annual conference. An attempt at decentralization would have been a direct attack on the trade unions' political power. All along the way, the CLPD had been very sensible and sensitive to the question of alliances with the trade unions; as one of its leading members put it: "Constituency Labour Parties have less than 10% of the more than seven million votes at party conference. The only way CLPD has won majorities at successive conferences has been by capturing the trade union block vote, emphasizing that we are fighting for the implementation of union policies, and *avoiding at all costs any action which our opponents could portray as an attack on the trade unions*" (emphasis added) (Koelble 1987, 260). A battle around decentralization would have been lost because the unions would have voted against it.

Figure 4.2 can help us visualize the CLPD's strategic choices. Consider the situation in the 1960s. In a schematic way, there is a coalition between the parliamentary group and the trade unions, players 1 and 2 of Figure 4.2A. In Brighton, the activists introduced a new issue: constitutional reform. This reform could be either decentralization (no control of the nomination process by the central authorities of the party) or mandatory reselection. In the first case, the trade unions (player 2 in Figure 4.2) would side with the parliamentary group (Figure 4.2B), and the reform would be defeated; in the second case, the trade unions (player 2) would side with the activists (player 3), and the outcome, as shown in Figure 4.2C, would be a new deal institution.

The fact that the essential modifications in the selection process were made by a change in majorities and not by a change of institutions is corroborated by the history of the schism between Labour and the SDP. After the annual conference in Wembley in January 1981, the Council for Social Democracy was created and two months later (March 1981) became the Social Democratic party. However, even before March 1981, the disagreement between the future leaders of the SDP and the Labour party over policies was not secret. "On 1 August David Owen, Shirley Williams and William Rodgers published a letter in the Guardian declaring their belief in policies almost wholly contradictory to the general trend of constituency opinion" (Kogan and Kogan 1982, 69). Williams (1983, 46) claims that party leaders arrived at their decision after growing discontent and a long hesitation throughout the 1970s.

An insider's (and therefore, partisan) view of the process is provided by Bradley (1981, 55), who claims that "Roy Jenkins now wishes that he had backed Taverne and provoked a split in the Labour party in the early 1970s." However, the rightward shift of the party while in office during the 1974–79 period renewed the hopes of the Right. Bradley (1981, 63) notes that "The founding of the CLV (Campaign for Labour Victory) was a sign of the acute anxiety felt by many social democrats by 1977. Some of them were already privately beginning to regard a split from the Labour party as inevitable." His account makes clear that there was a progressive shift in the minds of the "Gang of Three" (Owen, Williams, and Rodgers), who initiated the SDP even before the electoral defeat of 1979.

Bradley concludes that the major shift in the Labour party was the changing majority of the NEC and the annual conference and not the change in institutions (although their impact on the crystallization of the balance of forces should not be underestimated). At the heart of this shifting majority was an exogenous change: the political realignment of the trade unions. Consequently, an alteration of the trade union alignment can reverse the situation to the previous political predicament. After three consecutive Conservative party victories, there are signs that the trade unions realize the importance of their role in a new transformation of the Labour party.

IV. Conclusions

The chapter began with a puzzle: why do Labour activists replace their MPs and lead their party to electoral defeat? To investigate the answer, I developed a model of MP reselection. This model and its variations indicated that under perfect information and in a single-shot game, there is no possibility for any open conflict, not only between MPs and constituencies, but between constituencies and the NEC and the NEC and the annual conference as well. In addition, it is always the last actor with veto power who imposes her will. However, if the game is iterated and information is incomplete, it becomes rational for activists to reject their moderate MPs even if they prefer one additional Labour MP over a Conservative one. In the long run, they create a reputation for being tough and can teach their representatives to follow more closely the GMC's will. Similarly, conflicts between constituencies and the NEC can be interpreted as signals of a desire to shift the balance of political forces to the left or to the right. This reselection game is nested inside a competitive game between parties at the constituency and national levels. This competitive game modifies some of the actors' payoffs and increases or decreases the likelihood of certain strategies:

—A close race in a constituency strengthens the position of the moderate MP. A close race at the national level strengthens the position of the moderates inside the party while strengthening the importance of each individual constituency, which is able to blackmail the central leadership of the party.

—Political congruence between any actor and an actor with veto power increases the first actor's bargaining potential; that is why MPs deviating from the party line to moderate positions are rejected by their constituencies without NEC interference.

—The balance of power is influenced by institutional arrangements. The strength or the influence of different actors cannot be assessed by the frequency of disagreements or by the frequency of particular outcomes, because open conflict between actors in this particular game indicates neither weakness nor strength.

The analysis of the changes the CLPD introduced in the Labour party constitution in 1979 was shown to be much less important than is usually suggested in the literature. It was shown that the change in the political position of the trade unions made these changes possible, in addition to being the direct cause for the political shift of the Labour party.

Finally, the activists chose mandatory reselection as opposed to some more drastic institutional reforms (decentralization) for strategic reasons: to gain trade union support. By choosing to dispute control at the central level, CLPD leaders were able to win the whole party in the early 1980s but are losing it again as the majority of the NEC and the annual conference shifts to the right. In 1982, the NEC was recaptured by the Right, the shadow cabinet was largely right-wing, and the official persecution of the left-wing militant tendency began.

In a more general sense, this chapter deals with the question of motivated irrationality. The activists were choosing what would have been a suicidal strategy in a one-shot game because they were involved in an iterated and nested game. By choosing to replace their moderate MPs even when it had catastrophic consequences for the party, as the examples in the introduction to this chapter demonstrated, they were sending a signal to other candidates and to the leadership of the party: they could not count on their loyalty to the party and the structure of the reselection game to impose on them what they considered to be unacceptable solutions. Each time activists rejected a standing MP, they were sending an additional signal and creating a reputation of "toughness" or "irrationality" or, to use Webb's term, "fanaticism." In the future, they could cash in on this reputation because MPs would not dare to be moderate or because the NEC would think twice before engaging in a battle with the GMC.

The phenomenon of motivated irrationality is frequent in politics. Consider leaders like Qaddafi or Khomeini. They appear to be irrational not only by Western standards, but by practically any possible payoff matrix. Their behavior cannot be explained in terms of single-shot games. It can become intelligible *only* if one considers iterated games with incomplete information, where creating a certain image can be profitable in the long run.

In an even more general way, games like the one Labour activists play can help us understand the question of reputation building and its importance. An actor builds her reputation by choosing actions that seem suboptimal and can be explained only as the result of some particular characteristic. Keeping one's promises does not create a reputation of reliability when it is in one's interest to keep those promises. If, however, keeping one's promise will lead to an important loss, then a reputation for reliability can be built. Later on, this reputation, established on short-run sacrifices, can be used as an asset, just as activists use their reputation for irrationality as an asset.

Finally, problems where ideology is an important decision-making factor can be dealt with in a similar way. The traditional way of dealing with ideology in the rational-choice research program is exogenously, as an information cost-cutting device (Downs 1957). My approach suggests how to treat ideologies endogenously: they may have been adopted as solutions to recurring games. For example, in the short run, it may not make sense that Communists do not want to participate in coalition governments. If, however, one introduces long-term considerations, remaining faithful to the doctrine "do not manage the crisis of capitalism" may be an optimal strategy.

Appendix to Chapter 5:
Construction of Data Set Concerning Moderation and Marginality

The number of elected Labour MPs in the October 1974 election was 319. There were 13 replacements in the 1974–79 legislature. These additional Labour MPs are treated as independent observations, raising the number of observations to 332.

The Labour government lost twenty-two votes because of dissension of Labour MPs in 1974–79 (Norton 1980, 491–93). Of these twenty-two dissension-caused defeats, twelve concerned devolution bills. The remaining ten were the following:

1. January 29, 1975: government amendment to Social Security Bill

2. July 2, 1975: government amendment to clause 20 of Industry Bill

3. July 2, 1975: government amendment to delete schedule 3 of Industry Bill

4. July 17, 1975: Conservative amendment to Finance Bill on VAT on television sets adopted

5. August 4, 1975: government motion to disagree with Lords' amendment to delete clause 4 of the Housing Finance Bill

6. March 10, 1976: government motion on public expenditure

7. November 10, 1976: Lords' amendment to Dock Work Regulation Bill carried against the government

8. July 13, 1977: new clause to the Criminal Law Bill carried against the government

9. December 5, 1977: government defeated on adjournment motion following emergency debate on crown agents' affair
10. February 7, 1979: amendment to clause 4 of the Nurses, Mid-
    wives and Health Visitors Bill carried against the government

Of these ten cases, Norton reports the names of MPs voting
against the government or abstaining in nine cases (case 7 is missing). I used the sum of negative votes and abstentions in these nine
defeats to construct my dependent variable.

Chapter Six

A Rational-Choice Approach
to Consociationalism

The interaction between elites and masses in political decision making is an important issue in democratic theory. Crucial questions concern the importance of mass participation in the decision-making process and its consequences. On these questions, the literature is divided.

For the pluralist school, alternatively, elite competition for popular support defines democracy (Bentley 1908; Dahl 1956; elite is the primary mode of making the decisions" (Prewitt and Stone 1973, 152). As Mosca (1939, 156) put it, it is impossible for the masses "to exercise their right of option and control in any real or effective way." For Michels (1949, 166–69), the ruling elite is a "closed caste" that dominates society.

For the pluralist school, alternatively, elite competition for popular support defines democracy (Bentley 1908; Dahl 1956; Lindblom 1977; Truman 1951). In order to be elected, political leaders must fulfill the real or anticipated wishes of the electorate (Sartori 1978, 72–80). Competition among elites for public office makes their decisions responsive to the aspirations of the masses (Schumpeter 1947). The clearest demonstrations of this proposition are the so-called "Downsian" models of party competition, in which the tastes of the electorate are considered fixed and political parties try to adopt positions that will maximize their share of the vote.

A complete description of democratic decision-making processes must take into account both the horizontal channels of influence among the different elites and the vertical channels between the elites and the masses they represent. This chapter incorporates both the vertical and the horizontal channels of influence, presenting a general model of elite decision making in contexts in which the masses impose significant constraints on their representatives.

The chapter constructs a common framework to address three questions. First, under what conditions is it possible for political elites in segmented societies to pursue accommodating strategies, that is, strategies aiming to settle divisive issues when only a minimal consensus exists?[1] Second, if elites choose such strategies over long periods of time, why do their segmented followers continue to vote for them? The third question is, how can political institutions promote accommodating strategies?

The particular concerns of this chapter are patterns of conflict, accommodation, and institution building in consociational democracies. The arguments are theoretical, that is, conditional statements independent of particular temporal or spatial specifications. For reasons of expositional simplicity, I use examples from one country (Belgium), which presents the advantage of undergoing constant institutional remodeling.

The underlying assumption is that political elites participate in games in two different arenas: the parliamentary and the electoral. Each move they make has consequences in both arenas.[2] More precisely, political elites engage in a parliamentary game that is embedded or nested inside an electoral game.

Section I describes the Belgian case according to the consociational literature. However, considering elites either as independent, as the consociational literature does, or as simple representatives of the masses provides a poor description of the *interaction* between elites as well as of the interaction between elites and masses, that is, between the parliamentary and the electoral arenas. According to the literature on consociational democracies, elites are better off if they behave in an accommodating way. In consequence, the literature is not able to explain the phenomenon

1. For the definition of *accommodation*, see Lijphart (1968:103).
2. This idea is not new. Similar conceptualizations can be found in Machiavelli and Ostrogorski. More recent and relevant cases can be found in Fenno (1978); Fiorina (1974); Denzau, Riker, and Shepsle (1985); and in all the "retrospective voting" literature (Fiorina 1981; Key 1966; Kiewiet 1983).

of elite-initiated conflict. To account for such behavior, Section II introduces a game theoretic framework in which the divergent interests and evaluations of the political situation by elites and masses generate a game in multiple arenas, the parliamentary arena is connected to the electoral arena, and the situation in the electoral arena affects the payoffs of elites in the parliamentary arena. Section III applies the framework to cases of decision making and institution building in Belgium. In the conclusion, I discuss the advantages of this framework and its further applications.

I. "Accommodation" and Sophisticated Voting

When a political actor is confronted with a series of decisions presented sequentially, she can consider each either as an isolated event (an object of choice per se) or as part of a sequence of choices (an intermediate step toward a final outcome). In the first case, she chooses her most preferred alternative. This way of voting is called *sincere voting*. In the second case, she understands that the immediate question is immaterial; what matters is the choice of a path to arrive at the final outcome. This way of voting is called *strategic* or *sophisticated voting*. The reader not familiar with the concept may find it useful to refer back to the Finnish example in Chapter 1, where the Communists were shown to vote strategically to assure the election of their favorite cadidate, Kekkonnen.[3] I explore the insights generated by sophisticated voting and apply them to the study of consociational democracies.

The main theme of the literature on consociational democracies is the coexistence of "sharp plural divisions and close elite cooperation" (Lijphart 1977, 2). As Lijphart (1968, 103–104) puts it:

Dutch politics is a politics of accommodation. That is the secret of its success. The term accommodation is here used in the sense of settlement of divisive issues and conflicts where only a minimal consensus exists. . . . A key element of this conception is the lack of a comprehensive political consensus, but not a complete absence of consensus. . . . The second key requirement is that the *leaders* of the self-contained blocs

3. Another interesting application of sophisticated voting concerning the American Congress can be found in Riker (1983) and in Denzau, Riker, and Shepsle (1985).

must be particularly convinced of the desirability of preserving the system. And they must be willing and capable of bridging the gaps between the mutually isolated blocs and of resolving serious disputes in a largely nonconsensual context.

If I understand Lijphart correctly, the mutually isolated blocs in the population (Catholics, liberals, socialists) would like an intransigent position to be adopted by their representatives. However, the leaders understand that the whole system would eventually be destroyed if everyone remained steadfast. Therefore, they vote in a sophisticated way, taking not only the actual question into account, but also the long-term consequences of their repeated nonaccommodating voting.

Lijphart seems to suggest that political elites have genuine concern for the political system. But this motive for accommodating behavior is not the only one possible. In another article, Lijphart discusses "government by an elite cartel," which suggests more self-interested behavior on the part of political elites (in McRae 1974, 70–89). It is also possible that external threats or constraints explain such accommodating behavior. Cameron (1978) argues that it is more likely that partisan disputes will cease in the presence of an internationally competitive environment and an open economy. Katzenstein (1985) asserts that small countries (including those of the consociational type) do better in the international economy and improve their economic performance because they opt for a stable political environment. Another explanation would stress that elites practice accommodating behavior because they have longer time horizons than the masses (Axelrod 1984).

Although these explanations present different (but not mutually exclusive) motivations, they all account for the accommodating behavior of elites by invoking an interest of a higher order than the voters' demands. The common denominator of these explanations of consociationalism is that political actors vote against their immediate interests in order to secure more important interests in the long term, that is, they vote strategically. Such a conflict between long- and short-term interests is not uncommon. In fact, it has been argued that the choice of long-term instead of short-term interests is the most important characteristic of human behavior (Elster 1983; Shubik 1982, 63).

This account of consociationalism presents a series of problems. The first is that it focuses on elites and ignores followers. Is accommodation an acceptable strategy for followers? If so, why don't they change their positions? If not, why don't they replace their leaders? Short-term discrepancies between elite behavior and mass aspirations are not infrequent. After all, most governments have to make unpopular decisions upon occasion. However, such a discrepancy cannot exist for a long time, especially if issues are considered important. Elites have to explain their behavior and persuade the masses or they will be replaced by more competitive rivals.[4]

The second problem is that although consociationalism and sophisticated voting can account for elite accommodation, they focus on elite decisions and leave out intraelite strategy. This omission of the strategic aspect of elite behavior generates problems on both the theoretical and empirical levels. The theoretical problem is that the behavior of one elite is independent of that of other elites. It is always better for each elite to vote in a sophisticated way, at least from the point of view of policy-making and implementation. If accommodation is the outcome of sophisticated voting, then the optimal behavior for each elite is to accommodate regardless of what the other elites do. One can imagine cases, however, in which intransigence would be a better solution than accommodation, for example, if one knows that her opponent will adopt an accommodating strategy. Thus, accommodation cannot be the unconditionally best option for all elites.

The empirical problem with the sophisticated voting and the consociational literature is that elites are never expected to initiate conflict on their own, although they may resort to it if forced by their followers. However, other literature suggests that such elite-initiated conflict does occasionally occur. For example, De Ridder, Peterson, and Wirth (1978, 101) argue that for all sources of division in the Belgian political system, "rather than the issues deriving from the cleavages, the cleavages are invoked or partially mobilized to generate support for an issue arising from other sources of political competition."[5]

In Section II, I present a model to address these problems and to provide a more adequate framework for the study of consociational democracies.

---

4. This is one of the major criticisms of consociational theories. In fact, Keech (1972) and Barry (1975a, 1975b), among others, have argued that in several of these countries, followers do not seem as polarized as the consociational literature would predict.

5. See also Covell (1981).

II. Nested Games: The Electoral and the Parliamentary Arenas

According to the consociational literature, in consociational democracies, society is organized in segments or pillars, followers are polarized, and elites demonstrate accommodating behavior (Lehmbruch 1974; Lijphart 1969, 1977; Lorwin 1971; McRae 1974; Steiner 1974). To capture this difference in preferences, I employ the following model. Each segment of the population and its representatives must choose between two different strategies: to compromise with the other parties (C) or to be intransigent (I). The choices available to elites and masses are the same, but their respective preference orderings for outcomes differ. Table 6.1 is essentially a replica of Table 3.1 with different names for the strategies. It presents the differences between leaders and followers according to the ordering of outcomes. For simplicity, only two actors are considered.

The most preferred outcome for followers (who are polarized) is to be intransigent when the other players compromise. Following the terminology of Chapter 3, I call this payoff T (for temptation). Followers consider the converse situation (they compromise while everybody else is intransigent) the worst possible outcome. Call this payoff S (for sucker). The other two symmetric outcomes of mutual compromise (R for reward) or mutual intransigence (P for penalty) can be ordered in any way. If mutual intransigence is preferred, the game is deadlock (Abrams 1980). If mutual cooperation is preferred, it is a prisoners' dilemma.

For elites, as for the followers, the most preferred outcome is to be intransigent when the opponent compromises. The second best outcome is mutual compromise. Finally, yielding to an intransigent opponent and avoiding conflict is preferred over mutual intransigence. This fear of mutual intransigence is what distinguishes elite preference orderings from those of the masses. So for elites the game is chicken.

These simple games faithfully represent the behavior attributed to leaders and followers by the consociational literature. For the followers, a dominant strategy of intransigence exists. Regardless of whether the game is deadlock or prisoners' dilemma, and re-

TABLE 6.1. *Payoffs of possible games between elites.*

	C(ompromise)	I(ntransigence)
C(ompromise)	R_1, R_2	S_1, T_2
I(ntransigence)	T_1, S_2	P_1, P_2

$T_i > P_i > R_i > S_i$: Deadlock
$T_i > R_i > P_i > S_i$: Prisoners' dilemma
$T_i > R_i > S_i > P_i$: Chicken

gardless of the strategy pursued by their opponent, the followers' dominant strategy is to avoid compromise. Leaders are afraid of the consequences of mutual intransigence and prefer mutual compromise. However, if one group demonstrates intransigence, the other will acquiesce.

How would elites interact in this situation? In particular, why would they take their followers' preferences into account when they play with each other? In Chapter 5, I gave a complete description of the game between followers (the activists) and leaders (the MPs) in the case of the British Labour party. The crux is that leaders can be replaced if they do not promote the policies their followers advocate. In consociational democracies, the mechanism of leader selection does not operate so much at the general electoral level as inside each political segment, party, or pillar of the society, where competitive elites can replace the leaders who do not conform to their followers' expectations. We also saw in Chapter 5 that the actual replacement of the leaders is not necessary in order to understand that the electoral constraint is operative.

So the leaders must take their followers' preferences into account because of the existence of the electoral arena. But their own preferences, defined in the parliamentary arena, hold mutual intransigence as the least preferred outcome. Inequalities (6.1), (6.2), and (6.3) present the order of preferences in the three possible games:

Electoral arena $\quad T_{ei} > P_{ei} > R_{ei} > S_{ei}$ (deadlock) $\qquad$ (6.1)

$\qquad\qquad\qquad T_{ei} > R_{ei} > P_{ei} > S_{ei}$ (prisoners' dilemma) (6.2)

Parliamentary arena $\quad T_{pi} > R_{pi} > S_{pi} > P_{pi}$ (chicken) $\qquad$ (6.3)

The subscripts e and p stand for the electoral and parliamentary arena, respectively, and the subscript i refers to the parties or groups participating in the game.

There is an important difference between modelling elite behavior as the outcome of a game and considering it as a simple case of sophisticated voting. My choice to use game theory captures explicitly the phenomenon of interaction among elites. Although in sophisticated voting, one actor can outguess the others and use foresight to promote his interests, in game theory, the different opponents cannot take each other's behavior for granted. They make their choices in an environment in which outcomes depend not only on their own strategies, but on others' choices as well.

To make this difference in explanatory power clear, consider the game in the parliamentary arena. According to the sophisticated voting literature, each elite has an unambiguously best choice—to accommodate. According to my game theoretic representation, however, in the game of chicken (equation 6.3), as Chapter 3 indicates, each player's best choice depends on her opponent's choice (she should choose to accommodate if the opponent is intransigent and be intransigent if he accommodates). Later in this chapter, I show that the case of elite-initiated conflict follows as a direct consequence of this modelling choice, whereas it cannot be explained by the sophisticated voting literature.

To understand how elites play the nested game, consider two extreme cases. In one, the leaders play the game in the parliamentary arena, so they play chicken (described by inequality [6.3]). In the second, the leaders represent their followers' aspirations faithfully and play the game in the electoral arena, so they play deadlock (inequality [6.1]) or prisoners' dilemma (inequality [6.2]). In reality, leaders are interested in both arenas. As a result, their actual payoffs will be a convex combination of the payoffs in the two arenas. I choose the linear combination because of its simplicity. In algebraic terms:

$$PO_i = kPO_{ei} + (1 - k) PO_{pi} \qquad (6.4)$$

where PO_i stands for the payoffs (T, R, S, or P) of player i, and k is in the [0, 1] interval and indicates the weight of the electoral arena or the weight of the masses in the decisions of the leaders ([1 − k] indicates the weight of the parliamentary arena). If the masses care a lot about an issue, then the value of k increases, and the margins of maneuver of elites decrease. In formula (6.4), I have replaced the whole discussion found in Chapter 5 by the simple parameter k. In the new game, the players are political elites, and their payoffs are given by a linear combination of the payoffs of the parliamentary and the electoral arenas.

Equation (6.4) can generate three different orders of payoffs for each player. They are given by inequalities (6.1), (6.2), and (6.3). However, players need not weight the two arenas the same way (have the same value of k); therefore, they can rank their payoffs differently. Thus, there are nine (3 × 3) possible different nested games that can be played by two players. I return to this observation when I discuss the case of elite-initiated conflict. This simple idea draws on and can clarify Sartori's (1976, 143) distinction between visible and invisible politics:

In a first and uninteresting sense, a large slice of the political process escapes visibility because it is too minute and because we cannot keep our searchlights on everything. In a second sense, invisible politics is deliberately hidden and consists of its unpleasant and corrupt part: political money, spoils, clienteles, and dirty deals. . . . We are . . . referred to a third way of dividing the visible from the invisible part of politics, according to which the former corresponds to the words and promises destined for the mass media, while the latter corresponds to the deals and words for mouth-to-ear consumption. This is the distinction that bears on our discussion.

If elites play only in the electoral arena, then the masses are influential in the decision-making process, and the game is prisoners' dilemma or deadlock. Dominant strategies exist, and choices are clear and unconditional. That is why visible politics—that is, politics designed to be watched (and approved) by the masses—has an ideological and polarized character, as Sartori claims. However, if elites play only in the parliamentary arena and can make their own decisions, then the game is chicken. Their choices are contingent upon the opponent's strategy; if the opponent is intransigent, one prefers to yield rather than confront him. Politics becomes more pragmatic. The difference between visible and invisible politics is connected to the relative weight of the electoral and the parliamentary arenas, and both are reflected in the value of the parameter k, which indicates the influence of the masses on the decision-making process.

How do elites weigh the parliamentary and the electoral arenas? In other words, which variables influence the value of k? I think that there are two crucial factors: information and monopoly of representation. I investigate the impact of both.

(1) *Information.* We have to discriminate between two cases: whether the masses have information about *what* elites are doing and whether the masses know *why* elites behave the way they do. The second is conditional upon the first.

If the masses know, understand, and sympathize with the reasons for elite behavior, their own payoff matrix may be modified to resemble the elite's matrix. The game becomes a game of chicken not only for the leaders, but also for the followers. If, however, the masses know what elites are doing while disagreeing with their policies, the degree of freedom that elites possess decreases at least to the extent that the masses control elite action. Finally, if information costs are high, elites will possess a substantial degree of freedom from mass control. Obviously, invisible politics will be easier when the issue is not publicly salient, when there is another important matter attracting people's attention, when the issue is so complicated that the public cannot understand it, or when it is shrouded in secrecy. Thus, issue salience and visibility limit the freedom of elites by increasing the value of k.

(2) *Monopoly of representation.* The relationship between the electorate and political elites can be conceptualized as in Chapter 5. One implicit assumption of the model was that there was an available pool of representatives out of which the followers could select the replacement of their standing MP. If there were no other possible representative (in the case of a hostile NEC, for example), rejecting the standing MP would not make sense for the constituency. In our model, there may be no alternative choice for the masses because elites have monopoly control and the electorate is not able to reward or punish the elites. Put differently, electoral competition is essential for democracy.

I develop this point further because it is crucial to understanding the way consociational democracies function. Let us concentrate on one segment (pillar) of a consociational democracy. According

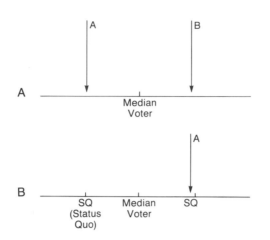

Figure 6.1A Competitive equilibrium in a one-dimensional space.
Figure 6.1B Equilibrium in a one-dimensional space with monopoly of agenda-setting power.

to the consociational literature, these segments do not communicate with each other, and the exit barriers for each are high. Consider a one-dimensional policy space, such as that presented in figure 6.1A. In addition, consider two different elites competing along this dimension and the position of the median voter. Under the conditions specified by Downs (essentially, the impossibility of abstention), the two competing elites will converge toward the position of the median voter. This holds *regardless* of the distribution of opinion inside the pillar. Indeed, regardless of whether followers' political opinions are unimodal or bimodal and regardless of the previously implemented policy (the status quo), electoral success inside the pillar implies convergence toward the position of the median voter. Therefore, competition between two elites is sufficient to drive them to the median follower's policy positions.

Compare this competitive situation with figure 6.1B, in which there is only one political elite. The masses either have to accept the elite's new proposals or remain in the previous situation. Now the position of the status quo becomes crucial because this particular political elite is in a position to "blackmail" its followers. The elite can propose anything in the SQ-SQ' interval. Because the followers must choose between this proposition and the previous status quo, any proposition in the SQ-SQ′ interval will be accepted. It follows that in a monopoly situation, the masses cannot impose their will. They have no choice but to accept a wide range of policies favored by the elite.

Factors that influence the availability of a rival elite are the salience of issues, costs for new elites to enter the electoral game (usually restrictions imposed by electoral law), and resources that the elite controls (existence of strong organizations and endorsements by other monopolistic organizations, such as the church).

There is evidence that in Belgium, a monopoly of representation was maintained on ideological grounds until the 1960s and on organizational grounds subsequently: Billiet (1984, 120, 123) claims that "there is a Catholic predominance in education and welfare—the two sectors that have expanded enormously in the last thirty years" and that "the pillar organizations are not only active as distribution channels, they are also involved in policy making and implementation." Huyse (1984) explains the process of *secular adaptation* of the different pillars by pointing out that the pillars developed new services (for the old and the disabled) or occupied activities that originated outside them (legal aid shops for the lower classes). The elites are thus largely unconstrained by the electoral arena, and they play chicken in the parliamentary arena.

Although the reasons for monopoly of representation inside each pillar can be described for Belgian society, the theoretical problem of monopoly representation and entry deterrence have not yet been solved. Existing attempts in the spatial voting literature indicate that the potential entrant's goal is of crucial importance.[6] Presumably, the existing elite(s) will attempt to prevent entry by adopting a position that will discourage challengers. If potential challengers want to become the most important representatives of the pillar, they can easily be discouraged because existing elites can position themselves in such a way that no entrant will become more popular than themselves. If, however, challengers are interested simply in acting as a blackmail group, their entry cannot be deterred by political maneuvers but rather can be prevented only by institutional constraints both inside parties and

6. For a literature review, see Shepsle and Cohen (1988).

at the general electoral level. Such constraints are provided by plurality electoral systems or thresholds for representation in proportional systems.

I do not claim to have adequately explained how a monopoly of representation can be achieved. The problem is crucial because an independent determination of k is necessary to understand exactly how elites weigh the two arenas and to perform systematic empirical tests of my model.[7] Still, we have a sufficient foothold for an initial application.

To summarize the argument: if elites enjoy a monopoly of representation inside the pillar *or* if information costs regarding elite behavior are high, the value of k is low; then elites are less constrained by the electoral arena and play a chicken game. If there is elite competition inside the pillar *and* information costs are low, the value of k is high; then elites have to conform to the demands of the masses, and a prisoners' dilemma or a deadlock game results.

Having examined the factors that influence the value of k (the influence of the masses in the decision-making process), one more important point requires attention before we apply the model to Belgian politics: whether the actors play a single-shot or an iterated game.

The most appropriate modelling simplification for how elites confront an important issue is a single-shot game. The reason is that for an important issue, the stakes are very high, so they cannot be traded off against other issues or promises about future behavior. In this case, as I have shown in Chapter 3, the decisive factor in determining the choice of strategies is the ordering of payoffs, not their actual values. The games described in this section as well as the games generated by any combination of payoffs described by (6.1), (6.2), and (6.3) have thoroughly studied solutions in pure strategies. So in the case of single-shot games, the conclusions of Section I of Chapter 3 can be readily applied.

If, however, the issues are of minor importance, trade-offs across issues can be effected. As a result, the situation is better

7. Moreover, one could argue that information costs depend on whether a monopoly of representation exists. Vigorous competition among rival politicians will reduce information costs and ensure that the relevant information reaches the masses.

approximated by an iterated game. As we have seen in Chapter 3, the value of the payoffs will influence the choice of strategies in iterated games (or in games in which correlated strategies are possible). Generally, as propositions 3.6 and 3.7 indicate, an increase in T or P makes the choice of defection more likely, but an increase in R or S makes the choice of cooperation more likely, regardless of whether the constituent game is prisoners' dilemma or chicken.

The relationship between payoffs and strategies indicates that we could predict (in a probabilistic way) the behavior of political elites if some evaluations could be made regarding the payoffs and the value of k involved in each case. Or, less ambitiously and more realistically, some comparative statics statements could be made about what type of behavior would be more likely under what conditions.

To summarize several important points: political elites have differential capacities to engage in parliamentary games; their payoffs in the parliamentary arena are set by the electoral arena; information costs and monopoly of representation are crucial conditions that determine those payoffs; as information costs increase, the value of k decreases; politics then becomes invisible, and the game between the different elites resembles chicken; as monopoly of representation increases, elites are able to choose their own policies, ignoring the desires of their followers. Regardless of the value of k, the game between different elites can be single-shot or iterated, according to the salience of the issue. If the game is single-shot, only the ranking of payoffs determines the optimal strategies. If the game is iterated, calculations over time or across issues are possible, rendering actual payoff values important. In the case of iterated games, the likelihood of cooperation increases when R or S increases while T or P decreases. I now use these intuitions to examine Belgian politics.

III. Studies in Belgian Politics

Today Belgium is a country with three linguistic groups (French, Dutch, and German); the constitution recognizes two cultural communities (French and Flemish), three geographic regions (Wallonia, Flanders, and Brussels), which do not coincide with the cultural communities, and a central government; there are three political families (Catholics, liberals, and socialists), which have created a dense network of social and economic organizations and institutions (Borella 1984).

The three political families represent the three "pillars" of Belgian society, according to the consociational literature. This means that the three political parties were able to monopolize representation of the corresponding segments of the population. However, between 1958 and 1961, territorial divisions became important. The population in Flanders exceeded the French-speaking population, and industrialization progressed rapidly in Flanders while Wallonia went into an economic crisis. From 1968 to 1978, the unity of the three Belgian political families was severely tested by the territorial question: in 1968, there was a schism between the Flemish and the French-speaking Christians (Social Christians party [PSC] and Christian People's party [CVP]). In 1972, the liberals were divided (Liberal Reform party [PRL] and Freedom and Progress party [PVV]). In 1978, it was the turn of the Socialists (PSB [for the French] and BSP [for the Flemish]) (Mabille 1986, 328).

In this section, I use the model generated in Section II in four different cases to explain Belgian politics: (1) to examine the case of elite-initiated conflict, (2) to examine the design of consociational institutions as an iterated game, (3) to study a particular historical event (the Egmont Pact) as a single-shot game, and (4) to study some findings of the sociological literature about cleavages.

1. Elite-Initiated Conflict

My account so far may have left the impression that because the game in the parliamentary arena is influenced by the electoral game, elites who are willing to accommodate (for the common good or for any other reason) are prevented from doing so by the masses. This is the general theme of the consociational literature; however, it is an account that has recently come under attack. Many political scientists now believe that political elites play important roles in creating conflict and in mobilizing the masses for particularistic interests. The phenomenon of elite-initiated conflict has been analyzed by other literatures. Schattschneider (1960, 5) writes of politicians trying "to re-allocate power by managing the

scope of conflict." For Riker (1983), the introduction of new
issues that divide the winning coalition and create possibilities for
new coalitions is the essence of politics. Bates (1974, 1982) and
Sklar (1963, 1979) have written extensively on the phenomenon
of elite-initiated conflict in Africa. Sabel (1981) analyzes why trade
union leadership groups may be intransigent. An example of elite-
initiated conflict is the mass mobilizations over linguistic issues in
the early 1960s, particularly the case of Flemish mobilization regarding the status of Brussels in 1961 (De Ridder and Fraga 1986,
378). The Flemish were much better organized and mobilized than
the Walloons and were able to force the issue onto the national
political agenda. Consociational theories, however, not only fail to
explain this type of elite behavior, but even fail to acknowledge its
existence (Covell 1981). Can the model developed here account
for such elite-generated conflict in consociational settings?

Political elites can mobilize their followers in order to increase
the share of the segment they represent in the parliamentary game.
Because the parliamentary game is one of chicken, it follows that
one of the players can force the equilibrium point to be (T_1, S_2) if
she commits herself first to the strategy of intransigence. She can
choose an issue that is both important to her followers (high T_1)
and unimportant to her opponents (high S_2). She can then mobilize
her followers in order to show the other parties that she means
business. Or she can use this political maneuver to discourage or
eliminate potential rivals inside her own pillar. In fact, at this
point, she can explain to her followers that she represents them
faithfully while convincing her opponents that she has lost control
of the situation and that the other party must capitulate. If the
issue is well chosen, she will have her way and will receive credit
from her followers. She can use this credit in the next round when
she receives an ultimatum from her opponent and has to capitulate. If such issues of asymmetric interest do not exist, conflict will
lead to the mutually feared outcome (P_1, P_2).

Alternatively, the value of k for one segment of the population
may be so high that the preference ordering of the corresponding
elite in the nested game is that of prisoners' dilemma, while the
preference ordering of the other elite may still be that of chicken.
In this case, one elite has a dominant strategy of intransigence
while the other has to yield and compromise.

2. *Institutional Design*

Let us study institutional design as a case in which the nested game (parliamentary, with constraints from the electoral arena) is iterated. This game is either prisoners' dilemma or chicken, depending on the value of k.[8] In both cases, there are important problems of coordination: in the absence of coordination, both players could end up with undesirable outcomes. However, if the two players could coordinate their activities, what would be the most desirable outcome?

The immediate answer is mutual cooperation: (R_1, R_2). Nonetheless, imagine a chicken game with the following (symmetric) payoffs: $T_i = 6$, $R_i = 3$, $S_i = 2$, $P_i = 0$.[9] If the two opponents played this game twice and chose to cooperate, their payoff would be six. If, however, they agreed to take turns defecting while their opponent cooperated, their payoff would be eight. We now see that the best strategy (assuming communication) depends on the actual payoffs. Figures 6.2A and 6.2B illustrate the optimal choices under different conditions.

Figures 6.2A and 6.2B represent each player's payoffs in a chicken game. In Figure 6.2A, mutual cooperation is the best outcome for both players; in Figure 6.2B, the best outcome is produced by alternating defection and cooperation with one's opponent. In technical terms, (R_1, R_2) is included in the Pareto set in the first case, but not in the second. In algebraic terms, the necessary and sufficient condition for the payoffs to be represented by Figure 6.2A is

$$R_2(T_1 - S_1) + R_1(T_2 - S_2) > T_1 T_2 - S_1 S_2 \qquad (6.5)^{[10]}$$

8. The only case in which this statement is not true is if the masses play a deadlock game (are extremely polarized), and the elites have no or very small margins of maneuver (existence of competitive elites and perfect information). However, even in this unlikely case, the qualitative results presented in this section remain unaltered.

9. Exactly the same argument can be made for prisoners' dilemma if $P_i = 2.5$. For convenience, I present the argument throughout for the game of chicken; however, readers should keep in mind that I refer to the general nested game, which might be chicken, prisoners' dilemma, or a game in which the two players have different payoff orderings.

10. This inequality is derived by writing the equation of the straight line through the points (T_1, S_2) and (S_1, T_2) and making sure that the point (R_1, R_2) lies to the northeast of it.

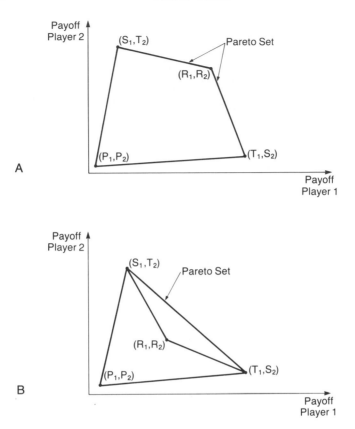

A

B

Figure 6.2A Chicken game in which mutual cooperation belongs to the Pareto set.

Figure 6.2B Chicken game in which mutual cooperation does not belong to the Pareto set.

It is easy to verify that as T_i or S_i increases, the probability that the point (R_1, R_2) will not be in the Pareto set increases.[11] As R_i increases, the probability that (R_1, R_2) is included in that set is increased. Alternatively, geometric intuition from Figures 6.2A and 6.2B suggests that as R_i increases, it becomes more likely that the point (R_1, R_2) will be above the line connecting the points (T_1, S_2) and (T_2, S_1), as in Figure 6.2A. Conversely, as T_i or S_i increases

11. This can be verified by checking the first derivatives of (6.5) with respect to the different payoffs.

(other things being equal), it becomes more likely that the point (R_1, R_2) will lie beneath the line connecting the points (T_1, S_2) and (T_2, S_1), as in Figure 6.2B. Moreover, as noted in Chapter 2, as T_i or P_i increases, the likelihood of the choice of a compromise strategy decreases. In addition, as R_i or S_i increases, the likelihood of compromise increases.

Let us now translate these game theoretic observations into political terms. When the payoffs R of mutual cooperation are high, the choice of cooperative strategies is both individually more rational and collectively optimal. In this case, it is easy to explain the cooperative behavior of elites because the benefits from mutual cooperation are high. Unfortunately, things are not always so straightforward.

Increasing T_i reflects the increased importance that a party attaches to having its own way on a particular issue. If the particular issue is not very important for the other party (high S), the likelihood of two events increases: the point (R_1, R_2) is no longer (Pareto) optimal, and the other party may give up on the specific issue. If, however, the same issue is salient for both parties, compromise is unlikely because a high T and low S increase the likelihood of intransigence on the part of both players.

The consequences are twofold. First, if an issue is very important for one party and less so for the other, two solutions are possible: either mutual cooperation or the second party gives in to the demands of the first. In an iterated game, the two parties can either choose mutual cooperation or take turns defecting and cooperating. Which solution is collectively better depends on whether (R_1, R_2) belongs to the Pareto set or not or, alternatively, whether (6.5) holds or not. But the evaluation is not easy to make, and political actors do not necessarily agree in their assessments. Second, on issues of mutual interest (high T and low S for both players), cooperation is most needed but most unlikely.

Let us consider issues that demonstrate the relevance of the payoff matrix. In Belgium, the Flemish community is traditionally very interested in maintaining its autonomy in education and cultural affairs; the Walloon community is interested in economic decentralization with respect to both investment decisions and expenditures. In our game, issues of great concern to one community (high T) are of low salience for the other (high S). Two solutions are therefore possible. If (R_1, R_2) belongs to the Pareto set, that is, if inequality (6.5) holds, the best solution is mutual cooperation, and both communities should decide both issues in common. If inequality (6.5) does not hold, the Walloon community should decide on issues of economic decentralization and the Flemish community, on cultural policies. One could imagine a tacit agreement in which the different groups would take turns making the decisions. This is the analysis of Billiet (1984, 124), who claims: "Not the Constitution, but a number of pacts determine the rules, consultation organs, and the consultation techniques." Dierickx's (1978, 144) account of Belgian conflict management is similar: "conflict can frequently be regulated with the help of package deals; it is profitable to make concessions where salience is low to obtain gains where salience is high." But how could such agreements and package deals be enforced? What would prevent a specific group and its representatives from shirking their obligations or claiming that the situation has changed, and they simply could not accept the agreement any longer? Such cases are possible and could destroy the cooperative intentions of elites.

Tacit agreements and package deals are possible and under certain conditions can provide nonconflictual solutions. However, there is no guarantee that such solutions will work because each group or coalition of groups can still claim control over the decision process on issues concerning other groups. A more efficient condition to avoid conflict over such situations is institutionalization of the decision procedure. Instead of deciding on conflictual policies, the different groups can decide on procedures about which their interests coincide. In particular, they can delegate authority to the most affected group.

So institutions can be designed to provide a permanent basis for the resolution of conflicts in which there is an asymmetry in the salience of issues for different groups under the condition that (R_1, R_2) does not belong to the Pareto set. This is essentially the account that Heisler (1973, 215) provides of the *dédoublement* process initiated in 1962 by the Lefevre-Spaak government: administrative separation was functional and preceded cultural autonomy. He claims that "*dédoublement* has helped to convert politically loaded issues into technically tractable matters." This is the situation if mutual compromise is not in the Pareto set.

More generally, the philosophical and political traditions of liberalism and federalism use the asymmetric importance of different issues for different groups of people. Instead of promoting a uniform solution that would be applied to the whole people or to the whole territory, liberal and federal solutions assign to different groups or territorial units the right to decide on issues of concern to them. The solution to the school problem is a demonstration of the creation of liberal institutions in Belgium. In 1951, the Social-Christians were alone in government and voted the Harmel law, which favored the Catholic schools. In 1955, when Socialists and Liberals came to power, they voted the Collard law, which imposed uniform standards on secular and Christian schools. In 1958, the Social-Christians won again in the election, but they signed an agreement with the leadership of the Socialist and Liberal parties that protected different philosophical beliefs, guaranteed the free choice of school, and provided government aid to all forms of education. This agreement became the law of May 29, 1959 (Mabille 1986, 322).

The institutional solutions adopted in Belgium include substituting the principle of territoriality for that of individual choice of language; mutual vetoes and special majorities; decentralization and group autonomy; "alarm bells" (*sonnettes d'alarme*), that is, control of the agenda by linguistic groups on issues of special interest; and the creation of special institutions to allow groups to monitor the execution of policies of special interest (Covell 1981). All these institutional arrangements share the characteristic of assigning priority over agenda setting or exclusive jurisdiction over issues either directly or indirectly (veto power).

However, we have not ruled out the possibility that the outcome of mutual compromise actually *is* in the Pareto set, nor should we. In fact, there are political actors in Belgium who believe that the best political solution lies within a unitary framework. Each political party houses unitary and regionalist tendencies. In particular, inside the Christian People's party, the two most important leaders and ex-prime ministers, Leo Tindemans and Wilfried Martens, disagree precisely on this issue. In terms of our model, the former believes that the best solution can be found through mutual concessions (that [6.5] holds). The latter promotes regionalist solutions. The differences between regionalists and their opponents is that the former believe they can do better through separation of issues and exclusive jurisdictions than through mutual compromise (that [6.5] does not hold), whereas the latter believe that issues are symmetric and that mutual compromise is better than a combination of unilateral decisions.

So far I have concentrated on three out of the four outcomes of the nested game. I have not discussed the outcome of mutual intransigence. The fear of this outcome makes elites desire compromise, and the lack of mutual compromise or coordination (in alternations of compromise and intransigence) can lead to it.

As demonstrated earlier, an increase in the value of T_i or a decrease in the value of S_i each reduces the likelihood that each player will choose to cooperate. Such conditions occur if an issue is important for both parties. In this case, mutual defection is the most likely outcome. If both parties could increase the payoff for mutual intransigence, this particular outcome would not be as painful.

The study of Belgian institutions indicates that efforts to increase the payoffs for mutual defection have been made and have been successful: when conflict seems inevitable, it is postponed. For example, article 59-b of the constitution requires that legislation concerning the composition and competence of community and regional bodies "must be passed with a majority vote within each linguistic group of both Houses, providing the majority of the members of each group are present and on condition that the total votes in favour in the two linguistic groups attain two-thirds of the votes cast" (Rudd 1986, 122). Moreover, any major constitutional revision, aside from qualified majorities, requires a statement by the government of articles to be revised, a dissolution of Parliament, new elections, and the formation and maintenance of a new government coalition.

The outcome of such stringent requirements for reforms is that some parliaments elected to perform constitutional revisions did not approve any: the legislatures of 1965–68, 1968–71, and 1978–81. Moreover, as a "Dossier du Centre de Recherche et d'Information Socio-Politique" reports, in 1983, the text of the constitution was incomplete: there was an article 107c (since 1980) and an article 107d (since 1970), but no article 107b (CRISP 1983, 5). Our game theoretic model, however, can provide a more interesting and less obvious reason for the lack of reforms than the difficulty of the requirements: as noted earlier, the fear of the payoffs from mutual intransigence (P) makes cooperation more likely. Indeed, as P decreases, the likelihood of cooperation increases. The institutional solutions adopted in the Belgian constitution, however, increase the value of P by postponing conflict, thus making cooperation more unlikely. So, paradoxically, the adoption of measures that reduce the consequences of disagreement (qualified majorities, postponement of conflict) increase the frequency of disagreement.

I have offered a rational account of consociational institutions. The reason for adopting such institutional devices, according to my account, is twofold: to push the outcomes of the nested game into the Pareto frontier (that is, to make outcomes collectively optimal) and to increase the payoffs of mutual intransigence by postponing conflict whenever it seems inevitable.

In my account, consociational institutions first *separate* issues that in other countries are usually connected, and they then *assign jurisdiction* to groups over issues of concern to them (Shepsle 1979). The account differs from the explanation of consociationalism through package deals offered by Billiet (1984) and Dierickx (1978). If package deals were enough to solve the problems, the development of conflict-reducing institutions would have been redundant. If Dierickx's and Billiet's accounts were correct, and package deals were sufficient, employing resources for consociational institution building would have been a suboptimal strategy. The very existence of consociational institutions indicates a fundamental mistrust between different groups and the necessity of enforcing mechanisms for conflict management.[12]

12. Readers might object that I have taken only questions of Pareto optimality—that is, collective rationality—into account, but that the essential problem concerns the conflict between individual (or particularistic) and collective rationality. In other words, although I have focused on the need to create institutions, I have not demonstrated how agreement over institutions is obtained. This objection is correct. However, as I have shown in Chapter 3, iterations or communication lead to any individually rational outcome; therefore, achieving the Pareto frontier is possible. Because elites are in continuous interaction, they are able to recognize when they are in a non-(Pareto) optimal situation, that is, a situation in which they need institutions. Consequently, they can correct their action.

3. The Egmont Pact (1977)

The Egmont Pact was an attempt by the major Belgian parties to create an institutional compromise to resolve the status of Brussels, which has been a conflictual issue between the Flemish and Walloon communities. Because Brussels is predominantly French speaking, if it became an independent area, Belgium would be divided into three regions, of which two would be French speaking. Otherwise, Belgium would be divided into one French- and one Dutch-speaking community. Because the issue is of major importance to both communities, concessions on it cannot be compensated by gains on other issues. According to the assumptions made in Section II, the Brussels issue must be examined as a single-shot game. According to Covell (1982, 457–58):

> The Egmont pact was negotiated in 1977 as part of the government formation process. The negotiating team included the potential prime minister and the presidents of the potential coalition parties. . . . It took place over a three week period, under conditions of secrecy and an intense pace that included several all-night sessions. . . . The isolation and secrecy with which the negotiations worked created what they describe as team spirit. It also aroused the suspicions of those excluded from the negotiations and created a situation in which the negotiators became more concerned with preserving their relationships with each other than with carrying their party organizations along with them. Each side came to believe that the main obstacle to an agreement was not the "adversary" with whom they were negotiating, but their own followers, who would have to be made to accept the agreements. In fact, the negative reactions of their followers were underestimated by all negotiators.

The pact could be implemented only by laws passed by Parliament. Leo Tindemans, then prime minister, was not in favor of the regionalist provisions in the pact and sought to delay its implementation. Moreover, the composition of Parliament was not favorable to implementing the Egmont Pact. In October 1978, Martens, president of the Christian People's party (CVP), moved negotiations back to the presidents of the parties. In this new round, some of the presidents threatened to initiate a governmental crisis over the lack of progress in implementing the Egmont Pact. Tindemans (CVP) then resigned, blocking the process even further.

Lijphart (1977, 182) thinks that the accommodating behavior of elites is an independent variable and can be used whenever needed. He states, "The more extreme the condition of cleavage and mutual isolation, the clearer the danger signals are likely to be perceived. Once the peril is recognized, remedies may be applied." Lijphart calls this a self-denying prophecy.

The independent status of elites in Lijphart's theory has been one of the major attractions of his version of consociationalism because, unlike versions advanced by Lorwin (1971), Lehmbruch (1974), and Steiner (1969), it lends itself prescriptively as a policy tool for elites to apply whenever they perceive the danger of cultural conflict.[13] However, the events surrounding the Egmont Pact do not corroborate a theory of independent elites: elites clearly cannot avoid conflict when issues are important for all parties.

Covell (1982) explains the negotiation phase of the pact as a prisoners' dilemma game and the implementation stage as a deadlock game. The problem with her explanation is that intransigence is the dominant strategy in both of these one-shot games. Thus, negotiations should have failed in the first place. However, her intuition that it matters whether elites negotiate on their own or the masses are involved in the process is essentially correct.

What intuitions about the Egmont Pact can we obtain from the model presented in Section II? The negotiations took place under extreme secrecy; therefore, the value of k approached zero in the negotiation phase. During the three weeks, the negotiators were playing a game of chicken in which the fear of failure and the perpetuation of the status quo concerning Brussels drove their decisions. Moreover, because they were communicating in the negotiations, they were able to develop contingent strategies and push the outcome toward the Pareto frontier. The possibility of communication and bargaining, that is, the possibility of contingent strategies, can lead a single-shot game between elites to the same outcome as an iterated game: to the Pareto frontier. In the implementation stage, agreements became public, and the value of k dramatically increased. Moreover, because the masses did not interact with each other or negotiate, contingent strategies were not

13. See also Halpern (1986).

possible any more. The game became prisoners' dilemma or dead-lock, so that intransigence was the dominant strategy. However, the parties' presidents, who are not directly connected with the voters (lower k), continued to perceive the game as chicken and used the ultimate threat (for a chicken game): to bring down the government. This, however, was not a threat to people like Tinde-mans, for whom intransigence was the dominant strategy and who therefore preferred the resignation of the government over the ratification of the Egmont Pact.

The model presented in Section II enabled me to explain impor-tant aspects of Belgian politics: the issue of elite-initiated conflict, the design of institutions, and historical events like the Egmont Pact. Alternative explanations do not consider or cannot account for these issues. As we saw, consociational theories and sophisti-cated voting theories fail to recognize or explain the existence of elite-initiated conflict. Moreover, other approaches (such as those of Dierickx and Billiet) that try to account for accommodation through package deals and cross-issue trade-offs fail to account for the specific form of Belgian institutions.

4. Segmentation

The same nested games approach can also be used to understand the importance of some sociological characteristics of consocia-tional democracies, such as segmentation. According to classic accounts, fragmentation is likely to produce political conflict (Almond 1956). However, if the various communities do not com-municate, it is likely that different political elites will exercise a monopoly of representation over their respective segments. Daal-der (1966, 214) describes Dutch society as presented by Kruijt, "the elder statesman of Verzuiling studies": people go to school; belong to unions, radio associations, and political parties; and read newspapers and books exclusively within the framework of their pillar (*zuilen*). If there is no competition *inside* the pillar, there is no competition at all. This situation gives political elites a large measure of freedom of action. If, however, there is internal competition, political elites will faithfully reflect the feelings of their electorate (see Section II).

In Northern Ireland, where political competition occurs within each segment, attempts at compromise inevitably fail because extremist leaders are able to mobilize the masses (Lijphart 1977; Schmitt 1974). One can compare the Irish case with the situation in the Netherlands as described by Daalder (1966), in which the established parties maintained their political strength. In addition, despite the absolutely proportional electoral system, challengers inside each pillar failed to receive more than 16 percent of the vote. The situation has completely changed since it was described by Daalder: political elites do not enjoy monopoly privileges any more because the secular trend in Dutch politics has increased the number of parties and the degree of electoral volatility (Thung, Peelen, and Kingmans 1982). If this secular trend had preserved the boundaries of the pillars, Dutch politics would have become very conflictual. What happened, however, is that the importance of the pillars themselves declined, so the reasons for conflict in Dutch society were eliminated.

In Belgium, political parties enjoyed a monopoly of representation until the ethnic issue became prominent. Compare the electoral failure of the new Union Démocratique Belge in 1946 with the subsequent success of regionalist parties (Lorwin 1966, 167) and the schism of all national parties into Flemish and Walloon groups (Heisler 1973).

According to my model, political elites who have lost their monopoly will accurately reflect the feelings of their constituents. However, these feelings also may have changed: the different communities may not be antagonistic any more. Thus, one cannot make a prediction in this case.

IV. Conclusions

Pluralist theorists argue that fragmented societies are doomed to political instability. Consociational theorists focus on elite political behavior, claiming that if elites compromise, the polity will be stable despite divisions at the mass level. These theories have come under attack for their conceptual fuzziness, static typological character, and discrepancy with the facts.

There are two crucial questions. Are there discrepancies between mass aspirations and elite behavior? Under what conditions and with what consequences? Differences between the conceptions of elites and masses can be explained as an indication of sophisticated voting rather than as a cultural phenomenon.

However, strategic voting is not able to capture contingent choices such as those made by political elites. The use of game theory allowed the distinction between single-shot and iterated games, which produce different outcomes. Moreover, iterated games were used to describe two different situations: when mutual compromise produces a Pareto optimal outcome and when the Pareto frontier is achieved only by alternating compromise and player intransigence. The first is the simpler case, and as long as elites enjoy a monopoly of representation, they can choose to compromise. The second case is more complicated and may require the use of constitutional devices for a Pareto optimal compromise to be implemented. Concerning issues of asymmetric importance, institutions assign exclusive jurisdictions and delegate complete authority to the concerned group. Concerning issues of symmetric importance, institutions minimize the consequences of disagreement by postponing conflict.

In the absence of specific institutions, elites might initiate political confrontation in order to signal that the issue is salient or to discourage potential rivals inside the segment. Consociational theories cannot explain this strategic use of conflict and mobilization.

I considered several sociological issues inside the framework of this model. What happens, for example, if the monopoly of representation is disputed within a pillar or if a pillar breaks down? In the first case, the outcome is extremely conflictual; in the second, the outcome is indeterminate.

This simple rational-actor model has the potential of combining theories such as consociationalism and Sartori's visible-invisible politics distinction. It can also account for phenomena related to the emergence of institutions and for elite-generated conflict, phenomena that until now were explained only by partial theories. In this sense, it is a very good demonstration of the translation and synthesis of different theories into a simple, empirically accurate, and theoretically fruitful rational-choice framework.

Chapter Seven

The Cohesion of French Electoral Coalitions

Coalition building involves both cooperation and competition, but the dynamics between these two elements has not yet been systematically analyzed. The existing game theoretic literature focuses exclusively on the cooperative aspect of participants in a government coalition (Axelrod 1970; Dodd 1976; Luebbert 1983; Riker 1962). The question posed by this literature is which coalition will form, not which coalitions (once formed) are likely to persist. Moreover, the zero-sum game assumption made either explicitly or implicitly by these authors implies that the coalitions formed will be of minimum size—an empirically inaccurate conclusion.[1]

Recognizing that cooperative and competitive strategies coexist inside an alliance implies that the cohesion of an alliance is itself a variable to be explained. To analyze this problem, I use the framework of nested games: I consider political parties as pursuing strategies in two different but connected arenas and their choices as affected not only by the balance of forces *between* coalitions, but also by the balance of forces *within* each coalition. The game between partners is nested inside the game between coalitions.

I consider the partners of each coalition to be playing a game

---

1. Grofman (1982, 86) presents a model of proto-coalition formation based on ideological proximity in which coalitions might not be of minimum size. His model, however, also assumes that "proto-coalitions, once formed, remain nondissolvable."

with variable payoffs. The payoffs vary according to the outcome of a (competitive) game between coalitions. So parties find themselves in a situation in which their payoffs vary according to the specific balance of forces between coalitions and have to choose strategies that have implications for the balance of forces both within and between coalitions.

Class conflict can be understood as a game between partners and coalitions because each social class confronts the other while facing its own collective action problem, and political influence ultimately depends on which class solves the collective action problem more effectively.[2] Primary elections in the United States present another case in which the same framework can be useful. Competition between candidates for party nominations may leave incurable wounds, thereby handicapping a party's chances of winning. Therefore, initiatives undertaken in the primaries have to be regarded (by both actors and observers) as having an impact on the general election. Finally, as we saw in Chapter 5, political tendencies or factions within political parties face games in multiple arenas: their decision to promote or undermine party unity has consequences on the competitive condition of the party.

This chapter focuses on the interaction between partners and between coalitions at the same time. In this chapter, the number of arenas increases. Besides the game at the national level, I consider the competitive game between coalitions and the game between partners. Moreover, in this chapter, I have sufficient data (on electoral outcomes) to concentrate on the empirical implications of the model and test the nested games approach. The framework is general enough to permit empirical tests, and for reasons that will become obvious, I have chosen the French elections of 1978 as the test case.

The chapter is organized into the following sections: in Section I, the reasons for the choice of France as a case study are given,

2. Or the balance of forces might be so favorable to one side that it does not need to overcome its collective action problem. Offe and Wiesenthal (1980) argue that this was the case for class conflict at the end of the last century: capitalists did not need to organize at the national level. Elster (1985, 346) and Przeworski (1985) provide evidence that the Marxian conception of class struggle can be captured by this formal approach because the force unifying each class is competition against another class. In other words, classes become classes against someone before they become classes for themselves.

and the possible outcomes of French elections are presented in a diagram that facilitates intuitive speculations about choices of party strategies. In Section II, the validity of these intuitions is examined through the theoretical framework of nested games. In Section III, empirical propositions derived from the theory of nested games are tested with French electoral data. In Section IV, an anomaly in the data leads to a distinction between visible and invisible politics and the laws that govern this distinction. In Section V, the conclusions of the nested games approach are compared with alternative explanations of French elections. In Section VI, the results of the previous sections are used to study the recent changes in French electoral laws (1985 and 1986) and demonstrate that they were the result of conscious institutional design by winners who wanted to consolidate their positions.

I. Why France?

The French Fifth Republic is an excellent case for studying coalition stability. Under the Fifth Republic, and at least up to 1984, the four major political families (the Gaullists, currently named the RPR; the Giscardians, currently called the UDF; the Socialists, called the PS since 1971; and the Communists, the PCF) formed two competing coalitions, the Right and the Left. The competition between the Left and the Right led to the progressive elimination (under the Fifth Republic) of center parties (Chapsal and Lancelot 1969). Duverger (1968) describes this system as *quadrille bipolaire* and explains that its mechanics are due to the particular electoral system used (with the exception of the 1986 elections) in the French Fifth Republic, namely, the two-round majority electoral system (from now on TRMES) in the National Assembly elections. In each constituency (*arrondissement*), each of the four major political families presents candidates for the first round of voting. If no candidate receives an absolute majority, then in the second round, held one week later, the party that came in second *within* each coalition usually endorses and supports the strongest candidate of the coalition (*désistement*). This intracoalition discipline is the result of agreements between the parties but is not enforced by the electoral law. The difficulty of sticking to this decision has resulted in cases of "triangular competition" (one candidate from one coalition and two candidates from the other competing against one another).

At the national level, the stability of French coalitions has been challenged several times:

—The Right moved from a period of Gaullist dominance (1958–74) through a slow reequilibration of forces under Valéry Giscard d'Estaing to ambivalent support for Giscard by the Gaullist party in 1981.
—The Left presented a single candidate in the first round of the presidential elections of 1965, split in the presidential elections of 1969, signed the common program of government in 1972, remained united until just before the legislative elections of 1978, when the common program was shattered, reunited for the elections of 1981 and in the first period of government (under Prime Minister Pierre Mauroy), only to split again in the summer of 1984 (after the Communist ministers withdrew from the government).

This history of conflict and cooperation is not unique. In all European democracies, parties join or leave coalition governments (the cases of the French Fourth Republic and Italy are the best known examples). What is unique to France is that both the cooperative *and* the competitive forces are magnified before the public because the electoral system favors both competition (in the first round) and cooperation (in the second round). As we shall see, a major advantage is that the visibility of the strategic maneuvering within or between coalitions offers opportunities for empirical research.

Duverger (1954) has demonstrated the implications of electoral laws for party systems. In proportional representation, the parties stress their differences to the electorate. After the election, government coalitions are formed, and the previous preelectoral competitiveness is replaced by cooperation within the government (at least as long as the coalition lasts). In plurality electoral systems, the two major parties try to build their electoral coalitions and reduce intraparty differences to the public as the elections approach.

In France, however, each party must do two things. It must affirm its own political line (otherwise it will lose its supporters in the first round); in the second round, however, it has to promote the coalition within which it seeks to capture control of the government. This situation is similar to the American primaries, which are followed by congressional or presidential elections. The important difference is that in the United States, a national convention or the simple passage of time *may* heal the wounds of the primaries.[3] In France, the two rounds are only seven days apart, so the parties do not have time to change their strategies. The simultaneity of elections as well as the visibility of strategies (coalitions are made *before* the election and *in front* of the electorate) make studying French politics especially advantageous.

If the two partners of a coalition go too far in criticizing each other in the first round, they will not have time to change their strategies in the second round and heal the wounds (even if they wish to). Some of the votes of the loser within each coalition will not be transferred to the winner; therefore, in the decisive second round, the coalition could lose. However, if a party is not sufficiently critical toward its partner in the first round, it might lose the decisive votes that would make it the front-runner in the first round and therefore give it the right to represent the coalition in the decisive second round.

Having set out the situation I seek to model, I now lay out the model itself. I begin by considering a single constituency represented in a particular space. This representation improves our understanding of the dynamics of cohesion and competition at the local level.

Ignore for the moment the internal divisions affecting the Right and the existence of smaller parties of both the Right and the Left, and consider the following (simplified) electoral competition: the Right (as a whole) confronts the two partners of the Left, the Socialists and the Communists. The equilateral triangle of Figure 7.1 can represent this triangular competition.[4]

3. Several times, however, the passage of time has not been enough to heal the wounds, and candidates do not endorse their fellow-party runners, or activists of a defeated candidate in the primaries join the other party (Johnson and Gibson 1974).

4. Figure 7.1 focuses on the internal divisions of the Left. To examine the Right, the dual triangular competition (between the Left, the Gaullists, and the Giscardians) would be relevant. Generally, the appropriate space to represent electoral outcomes is an n-dimensional Euclidean space (where n is the number of parties) and the corresponding n-1 dimensional simplex. The triangle of Figure 7.1 is in fact a two-dimensional simplex.

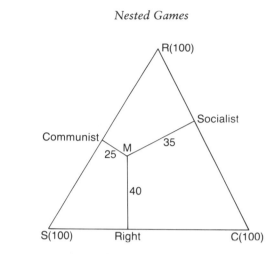

Figure 7.1 Two-dimensional simplex.

It can be shown that the sum of distances for any point inside the triangle to the sides of the triangle is equal to the altitude of the triangle. This geometric property can be used to map different electoral outcomes in a three-party contest on points inside the equilateral triangle. Each side of the triangle is named after a party (or coalition), and the distances of any point M from each side of the triangle represent the percentage of the vote of the corresponding party (or coalition). By definition (if we ignore other parties), these percentages sum to 100 percent. Setting the altitude of the triangle at one hundred produces a perfect correspondence between the percentage of the vote of a party and the distance from the corresponding side of the triangle. Figure 7.1 represents the electoral outcome in a constituency in which the Right coalition received 40 percent of the vote, the Socialists 35 percent, and the Communists the remaining 25 percent.

Figure 7.2 presents the same outcome space, but with some additional significant lines. C', S', and R' are the midpoints of the sides representing the Communists, the Socialists, and the Right, respectively. C'S' represents all the possible distributions of votes between Socialists and Communists that lead to ties between coalitions. Indeed, at any point of C'S', the Right receives 50 percent of the vote; the two parties of the Left, therefore, receive the remaining 50 percent. The segment GC' represents all the cases in which the following two conditions hold: the Socialists dominate the

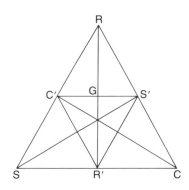

Figure 7.2 Ties between and within coalitions in a two-dimensional simplex.

Left, and the Right and the Left tie. The segment GS' represents the opposite case of a Communist-dominated Left. The vertical line RR' represents the ties within the Left. Along this line, the Communists and the Socialists receive the same percentage of votes. However, in the upper part of the segment (GR), the Left coalition is defeated; in the lower part (GR'), the Left wins the seat.

The area C'GR'S represents all electoral outcomes in which the Left wins and the Socialists are the stronger partner of the coalition. The area S'GR'C represents the case of a Communist-dominated and victorious Left. Within these areas, one has to distinguish between two cases: case A, in which one of the two coalition partners receives an absolute majority (triangles SC'R' and CS'R'), and case B, in which in order to win in the second round, one of the two partners needs the support of the other (triangle C'R'G for the case of the Socialists and S'R'G for the case of the Communists). Clearly, in such a situation, we may reasonably expect the weaker partner to possess considerable blackmail potential.

With respect to electoral outcomes, we can distinguish two sensitive zones: the vertical zone around the segment RR' and the horizontal zone around C'S', as Figure 7.3 indicates. Electoral outcomes anticipated to be inside the vertical zone are uncertain as to which one of the two partners will represent the Left in the second round. We would expect that in this area, the competitive aspect of party politics will win over the cooperative one. Electoral

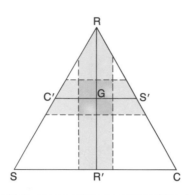

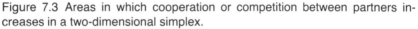

Figure 7.3 Areas in which cooperation or competition between partners increases in a two-dimensional simplex.

outcomes anticipated to be inside the horizontal zone of Figure 7.3 are uncertain as to which coalition will win. Therefore, the cooperative aspect of intracoalition politics is likely to dominate.[5] Note also that the nature of the competition is very different if the Left is expected to win a seat (lower part) or to lose one (upper part). In the former case, a seat is at stake; in the latter case, only an honorary title is at stake.

In summary, cooperation is likely when a seat is at stake (horizontal zone), but competition is likely when the two partners are almost equal in size (vertical zone). However, these geometrically generated political intuitions are incomplete in several ways. First, the two zones are not mutually exclusive; intuition will therefore be unreliable at the intersection of the two zones (the area around G), where each partner of the Left has approximately 25 percent of the vote. Second, the two zones are not collectively exhaustive of the outcome space; thus, for points outside the zones, we have no predictions at all. Third, the two zones are not defined in any theoretical or precise way; it is therefore difficult to say whether a point belongs to each one of them or not. Can we then deal with these circumstances within the same framework? It is to this question that we now turn.

5. An alternative hypothesis along the lines of Chapter 5, that is, that one of the parties would willingly accept a coalition defeat in order to improve its position in the long run, is examined in Section IV.

II. The Nested Games Approach

Let us now have a closer look at the game between the two partners of a coalition. Assume that each party has two alternative strategies: to cooperate with its partner (C) or to defect (D). By cooperation, I mean promoting the coalition's interests; by defection, I mean promoting partisan interests and openly criticizing the partner. Clearly, in the real world, coalition and partisan interests need not necessarily be in conflict, nor are parties restricted to two alternative strategies. For example, one can promote partisan interests without explicitly criticizing one's coalition partner or without attacking the partner directly or indirectly. The assumption that each party has only two strategies will be relaxed shortly. For the time being, however, let us examine the outcomes of this two-player game and try to imagine the payoffs for the two players.

A party (player) benefits most when it follows a partisan line while its partner promotes coalition interests (in terms of strategies, when it uses D while its partner uses C). This most preferred outcome for player i is called T_i (for temptation). The worst possible outcome is the reverse situation: when a party carries the weight of the coalition while the partner promotes its own interests. This is the intersection of strategy C with D of the opponent. This least preferred outcome is called S_i (for sucker).

The other two possible outcomes are mutual cooperation with payoffs R_i (for reward) and mutual defection with payoffs P_i (for penalty) for player i. We know that both these payoffs lie in the $[S_i, T_i]$ interval, but we do not know which of the two outcomes each player prefers. Disregarding ties, two orderings are possible:

$$T_i > P_i > R_i > S_i \qquad (7.1)$$
$$T_i > R_i > P_i > S_i \qquad (7.2)$$

If (7.1) describes the preferences of parties, then the game between the parties is a deadlock, and they would never form coalitions because when a party defects, it gets either the best or the second best outcome. If cooperation is to occur, the order described in (7.1) cannot hold.

If (7.2) holds, then the game between parties is a prisoners' dilemma. Each player is better off using a partisan strategy (no

TABLE 7.1.   *Payoffs of possible games between coalition partners.*

	C(ooperate)	D(efect)
C(ooperate)	R_1, R_2	S_1, T_2
D(efect)	T_1, S_2	P_1, P_2

$T_i > P_i > R_i > S_i$: Deadlock
$T_i > R_i > P_i > S_i$: Prisoners' dilemma
$R_i > T_i > P_i > S_i$: Assurance
$T_i > R_i > S_i > P_i$: Chicken

matter what the other party does), but if they both pursue this strategy, they find themselves worse off than if they had promoted the coalition.

Two more orderings of payoffs are possible and theoretically interesting:

$$T_i > R_i > S_i > P_i \tag{7.3}$$

$$R_i > T_i > P_i > S_i \tag{7.4}$$

If (7.3) holds, the game between the two coalition partners is chicken, and the worst possible outcome for each partner is mutual defection. If (7.4) holds, the game is an assurance game in which mutual cooperation produces the best possible outcome.

Table 7.1 replicates Table 3.1 and represents the game between partners at the national level. The preference order for each party is most likely given by (7.2). Indeed, it is more likely that each partner of each coalition prefers to criticize his partner and win votes, so the game is a prisoners' dilemma. Arguments can be made that criticism between partners is not unconditionally the best strategy. In this case, either (7.3) or (7.4) holds, and the game is chicken or assurance. If the game is considered one-shot and contingent strategies are not possible, as we saw in Section I of Chapter 3, then these modifications of the payoff matrix will produce different outcomes. If, however, contingent strategies are possible, as propositions 3.6 and 3.7 indicate, only the magnitude of different payoffs, not their order, determines the likelihood of the adoption of different strategies.

But this is not the only game. The game is subsumed within a competitive game between coalitions and a game between coalition partners at the constituency level. Parliamentary seats accrue to the strongest member of the winning coalition. Therefore, the incentives for cooperation or defection are modified by the electoral game at the constituency level. To determine the mechanics of these particular nested games at the constituency level, I proceed in the following way: (1) construct a new payoff matrix to take into account the utility of events at the constituency level (such as winning a seat or helping your partner win a seat), and (2) assess the impact of the new payoff matrix on the likelihood of cooperation.

1. The New Payoff Matrix

This matrix is constructed by adding to the payoffs of the original matrix at the national level (Table 7.1) the expected payoffs from the game at the constituency level. In order to calculate these expected payoffs, we have to define the utilities and the probabilities of different events at the constituency level.

Two probability distributions have to be defined over the space of electoral outcomes (the triangle in Figure 7.3). The probabilities p_v (v for victory) and p_{prox} (prox for proximity) are defined, respectively, as the probabilities that the anticipated outcome will be a tie between coalitions or a tie between partners. More precisely, p_v is an increasing function of the closeness of the outcome to a tie between coalitions (p_v is equal to 1 on the C'S' segment of Figure 7.3 and 0 on the segment SC and the point R). In algebraic terms:

$$\partial p_v / \partial \text{ victory} > 0 \qquad (7.5)^6$$

Similarly, p_{prox} is an increasing function of the closeness of the outcome to a tie between partners of the Left (p_{prox} is equal to 1 on the RR' segment of Figure 7.3 and 0 on the points C and S). In algebraic terms:

$$\partial p_{prox} / \partial \text{ proximity} > 0 \qquad (7.6)$$

Call V_i the utility of a seat to the coalition for party i. This utility will be different, depending on whether the seat goes to party i or to its partner. Call these two different possible values of V_i, W_i (for

---

6. ∂ is the sign for partial derivative. Inequality 7.5 says that the variable p_v increases as the variable victory increases.

win) and A_i (for ally winning), respectively. The values of W_i and A_i are an empirical matter. It seems reasonable, however, to assume that in all cases, $W_i > A_i$ because it is better for a self-interested player, such as a party, to win a seat than to leave it for its partner. Moreover, the value of A_i may be negative; a party might prefer its partner to lose the seat. Local rivalries or long-term considerations might account for such payoffs.

The expected value of a seat can now be calculated as the product of its utility (V_i) and the probability of winning it (p_v). In the case of a disputed seat, the victory can be assured only if both parties cooperate. In the case of competition, the stronger partner is likely to forgo the necessary transfer votes in the second round and thus lose. This reasoning suggests that the utility of mutual cooperation is higher at the local level than at the national level. More precisely, the expected utility of a seat has to be added to the utility of mutual cooperation. In algebraic terms:

$$R_i = R_i' + p_v V_i \qquad\qquad (7.7)$$

where R_i is the new utility (at the local level), R_i' is the utility at the national level, p_v is defined by (7.5), and V_i is either W_i or A_i.

The previous discussion concerns the dispute between coalitions for a parliamentary seat. What happens with the intracoalitional dispute? The crucial question concerning this dispute is who is to represent the coalition in the second round. Call U_i party i's utility from representing the coalition in the second round. If defeat is anticipated, this representation will have purely symbolic meaning. Call REP_i the value of U_i in this case. Representation of the coalition may, however, be of paramount importance when a seat is at stake. Call SE_i the value of U_i in the case of anticipated victory.

The value of SE_i is always positive and greater than REP_i because parties prefer to win seats. However, it is not clear, theoretically, whether the value of REP_i is positive or negative. Arguments can be made both ways. A party might prefer to represent the Left despite the probability of defeat because it thinks this would improve its position vis-à-vis its partner and, in the future, with a better balance of forces between coalitions, its probability of winning the seat. However, the party might also think that representing the Left when it loses is a liability for the future.

TABLE 7.2. *General payoff matrix for one coalition partner.*

	C(ooperate)	D(efect)
C(ooperate)	$R = R' + Vp_v$	$S = S' - Up_{prox}$
D(efect)	$T = T' + Up_{prox}$	P

Note: Payoffs are functions of the probability of a tie between coalitions (p_v) or a tie between partners (p_{prox}). V is the value of winning a seat by a party (W) or its ally (A). U is the value of representing the coalition when it is about to win (SE) or lose (REP).

The expected value of representation of a coalition can now be calculated as the product of its utility (U_i) and its probability (p_{prox}). This expected utility will modify the payoffs at the national level: if it is positive, it will increase the temptation to defect and decrease the sucker's payoff. Indeed, partners will have an additional incentive to be aggressive against each other if they can ensure themselves representation of the coalition and (maybe) a seat down the road. Conversely, being treated as a sucker will be more painful. In algebraic terms:

$$T_i = T_i' + p_{prox}U_i \qquad (7.8)$$

$$S_i = S_i' - p_{prox}U_i \qquad (7.9)$$

where T_i and S_i are the new utilities at the local level, T_i' and S_i' are the utilities at the national level, p_{prox} is defined by (7.6), and U_i is either REP_i or SE_i.

Table 7.2 represents the new payoff matrix for the nested game in each constituency. For reasons of simplification, only the payoffs of the row player are presented; therefore, the subscript i has been dropped.[7] The nature of the nested game represented by the new matrix is variable. For appropriate values of the different parameters, it can become an assurance game (in the area close to the segment C'S' of Figure 7.3 and for sufficiently high values of V), it can remain a prisoners' dilemma, or it can become a game of chicken (in the area close to the segment RR' of Figure 7.3 and for

7. It is, however, useful to remember that all parameters are indexed by party, and the value of an additional seat for Communists may be very different from that for Socialists. Consequently, all the comparative statements that follow concern the behavior of the *same* party (under different expected outcomes), not comparisons of different parties.

negative values of U). It can also become any other game in which the ordering of players' payoffs is not the same, and each player's ordering is given by any one of (7.2), (7.3), or (7.4). However, given that contingent strategies are possible and according to propositions 3.6 and 3.7, we are not concerned with changes in the nature of the constituent game, but only in the magnitude of each player's payoffs.

2. *The Cohesion of the Coalitions*

In Chapter 3, I demonstrated that in a prisoners' dilemma, chicken, or assurance game, the likelihood of cooperation increases when the payoffs for cooperation (R or S) increase and that it decreases as the payoffs for defection (T or P) increase.[8] Let us now examine the impact of variations of payoffs or distances from the lines RR' (tie between partners) and C'S' (tie between coalitions). We can distinguish the following cases:

V is negative. Negative V means that the value of winning an additional seat (W) or the value of one's partner winning an additional seat (A) is negative. Earlier we excluded the first but not the second possibility. If A is negative, then the closer an ally is to winning a seat, the higher the probability of winning (inequality [7.5]), and the more the reward from mutual cooperation (R in Table 7.2) decreases. However, the more R decreases, the more defection becomes an attractive strategy because its dominance becomes more pronounced. So if A is negative, that is, if for one party, the value of its partner winning a seat is negative, then the closer the coalition is to disputing the seat, the more likely the party is to undermine its partner.

V is positive. Similar reasoning for positive V indicates that coalition cohesion increases when victory is near. In particular, because for each party W > A, the dominant partner of a coalition will be more sensitive to the proximity to victory. We can summarize these results in the following proposition:

Proposition 7.1. When V is positive, coalition cohesion increases the closer the anticipated outcome is to a tie between coalitions; it decreases when V is negative.

---

8. Axelrod (1984, 202–3) and Maynard Smith (1982, 207–8) prove such propositions concerning the prisoners' dilemma game.

U is positive. Positive U means that the value of winning a seat (SE) or simply representing the Left (REP) is positive. I have provided arguments why this is always the case for SE and true most of the time for REP. It is always true that the closer the anticipated result is to a tie between partners, the higher the probability of a tie (inequality [7.6]). So as Table 7.2 indicates, if U is positive, the value of T (the temptation to defect) increases, and the value of S decreases (fear of being cheated increases). This means that the dominance of defection becomes more pronounced, making the choice of strategy D more likely.

U is negative. Similar reasoning for negative U indicates that coalition cohesion increases when the two partners are approximately equal. I have argued that this will occur if a party does not want to represent the Left when it is about to lose (REP < 0). The following proposition summarizes these results:

Proposition 7.2. When U is positive, the cohesion of a coalition decreases the closer the anticipated outcome is to a tie between partners; it increases when U is negative.

Taken together, propositions 7.1 and 7.2 indicate (1) that most of the time (except when the value of the victory of a seat by the ally is negative), coalition cohesion increases when the anticipated outcome is close to a tie between *coalitions* and (2) that most of the time (except when the value of representing the coalition when it is about to lose is negative), the cohesion of a coalition decreases when the anticipated outcome is close to a tie between *partners*.

The simplest algebraic representation of these two propositions is the following equation:

$$\text{cohesion} = c + (aV) \text{ victory} - (bU) \text{ proximity} \qquad (7.10)^9$$

where cohesion stands for the cohesion of the coalition, c is a constant, victory stands for the closeness of the anticipated outcome to a tie between coalitions, proximity stands for the closeness of the anticipated outcome to a tie between partners, V is the utility for a party of a seat that goes to the coalition, and U is the utility

---

9. Equation (7.10) can be formally derived as a Taylor series first order approximation of the likelihood of mutual cooperation (that is, cohesion) if one uses the chain rule because the signs of the required first derivatives are given in the text. This remark indicates that one could increase the precision of approximation and use nonlinear estimation routines for the empirical part. I do not follow this direction here.

for a party from representing the coalition in the second round. The appendix to this chapter gives the exact algebraic definition of these variables. The coefficients a and b are positive, as propositions 7.1 and 7.2 indicate.

A comparison of these conclusions with the intuitions proposed at the end of Section I indicates the following:

The epistemological status of propositions 7.1 and 7.2 and of equation (7.10) is different from the conclusions of Section I. Similar propositions were *conjectured* at the end of Section I; they are *derived* here from the nested games approach. Attention to this difference is not a statement of epistemological preference. Deriving propositions instead of positing them has the advantages of generality, better approximation, and specification of the conditions under which the propositions hold. I treat each advantage as a separate point.

Equation (7.10) does not concern French politics alone. It can cover cases of coalition cohesion such as those noted on pages 187–89, provided we can measure the independent variables.

Equation (7.10) covers the entire outcome space. We can therefore generate and test predictions about the intersection of the vertical and horizontal zones, as well as the areas not covered by the zones. In fact, the crude dichotomies generated by the two zones are now replaced by continua of outcomes. Moreover, calculus techniques permit us to replace the linear formula of equation (7.10) with more precise approximations.

Although my conjectures were largely correct, they were misleading on two points. This is another case where bare intuition, which may be an important guide to research, leads to incorrect conclusions that more rigorous formal reasoning is able to avoid. Cohesion does not always increase when the two coalitions are of equal strength. The condition for such behavior is that the weaker party in the coalition *wants* its partner to win the seat. This is neither a trivial assumption nor empirically correct in this case. Moreover, cohesion does not always decrease when the two partners are of equal strength. The condition for such behavior is that both parties want to represent the coalition *even when it is about to lose*. This, again, is not a trivial assumption, but it turns out to be empirically correct.

III. Testing for Cohesion

I use the results of the March 1978 elections for the French National Assembly to test equation (7.10). The reasons for choosing this year will become clear from a brief sketch of the relevant history of the French Fifth Republic.

From 1958 to 1974, the Gaullists dominated the Right, and the Right was in charge of the government. From 1974 to 1981, under the presidency of Giscard d'Estaing, a new balance of forces was created inside the Right, and Gaullist dominance was challenged. In fact, the UDF was created one month before the 1978 elections in order to mount a more effective electoral challenge to Gaullist dominance.

This period was also characterized by a change in the balance of forces within the Left when the new Socialist party created in Epinay in 1971 became the most popular party in France. The first national election in which the Socialist party became the most popular party in France and the dominant force inside the Left was that of 1978. Finally, although in 1978 the Left came close to winning, it held power only from 1981 to 1986.

This brief overview illustrates that the 1978 elections present two very important characteristics for this study:

(1) *The two coalitions were competitive.* The two coalitions were of almost equal strength in 1978, when the vote for the Left in the first round was 49.5 percent, compared with 46.3 percent in 1973 and 55.8 percent in 1981 (Wright 1983, 190). Because the two coalitions were of approximately equal size, one would expect, based on the theory just developed, maximum cohesion within coalitions.

(2) *The two coalitions were not cohesive.* Within both the Left and the Right, an important shift in the internal balance of power was taking place. The two political families of the Right competed widely in the first round, and the Socialist party demonstrated its dominance within the Left for the first time in 1978.

For these two reasons, both centripetal and centrifugal forces are expected to be more pronounced during the 1978 election. Thus, this particular election is especially appropriate as a test case for a theory of coalition cohesion. Therefore, I used the election results of the 474 constituencies of metropolitan France in 1978.[10]

Before proceeding to empirical tests, the variables indicated by the theory have to be operationalized in terms of the data. Two remarks are in order here. First, how do we operationalize the variable anticipated outcomes? I use the results of the first round as a proxy for this variable. This assumes that the parties have a fairly accurate perception of the electoral outcome, a legitimate assumption given the feedback from the electoral campaign that parties get both from their activists and the polls (which in France can be conducted but not published during the last week of the campaign). Once the anticipated result is equated with the actual result in the first round, the operationalization of the positioning variables victory and proximity is straightforward.

Second, how do we operationalize the variable cohesion? I have already argued that if a party does not cooperate with its partner but instead aggressively denounces its partner's policy positions, then even if this position is modified the day after the first round, its supporters will find it difficult to transfer their votes to the party considered their enemy only a few days earlier. Competition, therefore, results in the inefficient transfer of votes between the two partners in the second round. I use the difference between the votes of a coalition in the second round and the sum of the votes of the partners in the first round as the best indicator of the cohesion of the coalition.[11]

10. Overseas departments (DOM) and territories (TOM) are omitted.

11. This operationalization presents a problem in that it ignores vote transfers that do not appear on the aggregate level. For example, if the Socialist represents the Left in the second round, one cannot discriminate between the following cases: (1) all Communists transfer their votes and (2) some Communists abstain while some abstainers in the first round vote Socialist (or vote for the Right while some votes from the Right are transferred to the Socialist). Unfortunately, there is no way to correct for such ecological fallacies with aggregate data. However, because of the polarized electoral climate, I do not think that the "invisibility" of the aggregate transfers is very significant. Moreover, in this discussion, the interaction between party leadership, local party officials, and voters is ignored. In fact, the empirical outcomes may be attributed to strategies elaborated at the national or local levels, strategies followed precisely by the voters. Alternatively, they can be considered the result of independent decisions made by the voters themselves in the specific political environment. This does not mean that there is no strategic voting, that is, voting contrary to one's nominal preferences, because, as I show, parties (or voters) sometimes do not transfer *all* the votes to their partner (defective transfer of votes). More realistically, one could argue that dif-

One more point needs to be clarified. One might think that a coalition's maximum cohesion occurs when the votes in the second round are the same as the sum of the partners' votes in the first round. In this case, the partner delivers the coalition as many votes as it had in the first round. What happens, however, if the coalition gets more votes in the second round than it got in the first? This in fact happens frequently given that turnout increases by approximately two percentage points (Denis 1978, 981). But if turnout rises in the second round, this may be due to general factors (such as the perceived closeness or political significance of the result) rather than specifically local conditions. Thus, cohesion should account for the variance of vote transfers once this general increase in the second round is taken into account. Therefore, the constituency in which the coalition gains the highest percentage point increase in votes is the most cohesive. Note that this conceptualization of the problem leads to more conservative tests because transfer of all the first round votes to the representative of a coalition is no longer considered all a party can do for its partner.

This conceptualization of cohesion leads to the exclusion of several constituencies from the data set. First, it excludes all constituencies in which the winner was decided in the first round. Second, it excludes constituencies characterized by triangular competition (two candidates of the same coalition running in the second round). In this case, it would be inappropriate to sum the votes of candidates who run against each other. Third, it excludes constituencies in which only one candidate is represented in the second round. In this case, one of the two coalitions could not present a candidate in the second round (owing to the threshold restriction imposed by the electoral law) or would not (because it understood that there was no chance of winning); there is, therefore, no way to measure its cohesion. Of the 474 constituencies, 70 (15 percent) fall into one of these three categories. The first is

---

ferent parties have different levels of control over their voters and that this control increases *ceteris paribus* from Right to Left and from moderate to extreme parties. However, I choose to ignore this part of the interaction between voters and parties. In what follows, it does not matter whether vote transfers originate with party headquarters, local candidates, or the voters themselves. The reasons for this choice are the obvious simplifying consequences for modelling.

by far the most frequent: it includes 44 constituencies in which
there was a sole candidate of the Right who won in the first round.
Such cases are, in fact, cases of maximum cohesion of the Right in
which one of the two partners puts the coalition's interest over its
own. Such cases should therefore be included in the data set for the
Right and will be assigned the maximum cohesion score (which
turns out to be .091). But these cases cannot be included in the
data set for the Left because no indication of the cohesion of the
Left is given. Thus, my empirical investigation concerns 448 constituencies for the Right and only 404 constituencies for the Left.

   This accounting rule tips the scale in favor of my theory because
cases with lowest proximity are assigned highest cohesion (as the
theory predicts). I have two arguments in defense of my choice.
The first is theoretical. I have chosen to operationalize cohesion as
the difference of votes between two rounds because it was an
objective, easily available, and quantifiable indicator. The incon-
venient consequence is that cohesion's particular empirical referent is not defined in the absence of a second round. However,
regardless of the empirical referent of cohesion, cases where one
candidate withdraws in favor of his partner should be counted as
cases of maximum cohesion. The second argument is empirical.
The signs of the regression coefficients remain the same if the
forty-four constituencies are dropped out of the data set.

   For the convenience of readers, the equation to be tested, equation (7.10) is repeated here.

$$\text{cohesion} = c + (aV)\text{victory} - (bU)\text{proximity} \qquad (7.10)$$

Readers are also reminded that this equation was derived under
the simplifying assumption that the vote was divided into three
parts: the two partners of one coalition and the opposite (unified)
coalition. This simplification was necessary to introduce a two-
dimensional outcome space (the equilateral triangle) instead of an
$(n - 1)$-dimensional simplex. It is time now to relax this simplifying assumption and take the other parties into consideration.
Equation (7.10) indicates that the smaller the difference in size
between the two partners of a coalition, the weaker the cohesion
of the coalition (if U positive). In other words, the stronger the
second partner of a coalition, the less cohesive the coalition. Similar reasoning in a more complicated multidimensional space sug-

gests that other important allies reduce coalition cohesion in the same manner as one ally does. This reasoning indicates that an additional term expressing the strength of other allies has to be introduced into equation (7.10) for reasons of theoretical consistency.

$$\text{cohesion} = c + (aV)\text{victory} - (bU)\text{proximity} - (d)\text{others} \quad (7.11)$$

Examination of equation (7.11) indicates that it is the same as equation (7.10) with one additional term. This term is introduced to control for the importance of other allies in the coalition.

One improvement on (7.11) can be considered: the value of an additional seat is not the same regardless of the opponent's identity. For example, in France, where the Communist party was excluded from the political game for a long time and the right-wing parties deliberately used anti-Communist propaganda to undermine the Socialists, one would expect that the transfer of votes inside the Right would be much easier and more effective against a Communist than against a Socialist opponent. Similar results could be expected for the cohesion of the Left when its right-wing opponent was the RPR and its leader Jacques Chirac; both were considered very conservative. In fact, Jaffré (1980) reports survey evidence that corroborates the second conjecture but not the first.

Table 7.3 indicates the outcome of the estimation of equation (7.11) using the ordinary least squares (OLS) procedure.[12] The first line of Table 7.3 represents the results of estimation of equation (7.11) in the 141 cases in which the PCF represented the Left (and the PS had to transfer its votes in the second round). The table also presents the R^2 of the estimation with the values of coefficients written on top in each cell and t-statistics below in parentheses.

Out of twelve estimated coefficients (for victory, proximity, and others for each one of the four political families), one has a wrong sign and two are not significant at the .05 level (t < 2);.the

---

12. The additional dummy variable is added to (7.11) for the identity of the adversary. It might be argued that OLS is not appropriate in this case because the residuals may be correlated. However, using OLS does not bias the estimates, although it does decrease their efficiency, making hypothesis testing more conservative. So if OLS coefficients turn out to be statistically significant, this holds *a fortiori* for the GLS coefficients (Hanushek and Jackson 1977).

TABLE 7.3. Cohesion of French coalitions as a function of different variables.

Coalition	Repr	N	R^2	Cons	Victory	Prox	Adv.	Others
Left	PCF	141	.56	-.36 (-10)	.43 (10)	-.06 (-2.2)	.002 (.5)	-.36 (-5.8)
Left	PS	263	.09	-.07 (-1.9)	.06 (2.2)	.037 (1.55)	-.00 (-.4)	-.25 (-4.5)
Right	UDF	205	.44	.05 (1.1)	.09 (1.9)	-.10 (-6.7)	-.02 (-4.7)	-.35 (-8.1)
Right	RPR	243	.53	.07 (2.0)	.05 (1.4)	-.08 (-6.8)	-.02 (-6.3)	-.44 (-11.6)

Note: The different variables include their probability of winning, the distance between partners, the existence of other allies, and the identity of the adversary. (Adv. is considered to be RPR for the Left, and PS for the Right.)

remaining nine are significant at practically any confidence level. In three out of the four cases, the fit of the model is quite satisfactory (R^2 from .44 to .56). The only exception is the case of Communist party vote transfers, which produces both a very poor fit and the only coefficient with a wrong sign.[13]

What are we to make of the unsatisfactory fit of the model for the Communist voters? Why do Communist voters behave differently than supporters of other parties? To the student of French political life, this finding should not come as a surprise. The Communist party began a vigorous campaign against the Socialists in the summer of 1977 when the negotiations for the common program of the Left reached an impasse. During the entire electoral campaign, the Communist party refused to commit itself to the "discipline of the Left" because it considered the discussions of vote transfers premature and a distraction from the major issue, which was the negotiations for the common program (Lavau and Mossuz-Lavau 1980, 138). So the PCF's electoral strategy remained unknown literally until the last moment. It was *after* the first round (and only one week before the second), on March 13, that the three parties of the Left met and signed a vague political agreement that included vote transfers. This agreement was operative for only one week, and all partners denounced it after the second round.

Given the absence of clear strategy for PCF voters, it is not surprising that the vote transfers by Communist voters look like random noise, and the fit of the model to Communist behavior is poor. This is, however, only part of the explanation, and not the most interesting one.

IV. Visible and Invisible Politics

There is another way to explain Communist party electoral tactics. Let us try to reconstruct the political situation in the constituency immediately after the first round. One coalition has won more than 50 percent of the vote, and *if everything else remains equal and vote transfers work*, it will win the seat in the second round. In

---

13. Contrary to Jaffré (1980), the results indicate that the opponent does not make any difference for the cohesion of the Left, but that it does for the Right.

this case, the smaller partner of the winning coalition has the role of kingmaker: by transferring the votes, she can transform a good electoral performance into a parliamentary victory; by not transferring some of them, she can generate an electoral defeat. So the interesting political question between the two rounds is how well the smaller partner of the winning coalition will transfer its votes. Public attention is concentrated on the weaker party of the winning coalition. If the total votes of a coalition places it in the second position in the first round, then the excuse can be made that the coalition would lose anyhow, and public attention focuses on the vote transfers of the opponent's weak partner.

According to this reasoning, one would expect party supporters of the weaker partner of a coalition to run to the rescue of their partner (as Section II indicates) only when the combined votes of the coalition place it ahead in the first round, making visible the game of the expected winners in general and the minor partner in particular. In this case, the minor partner of the expected winner attracts the attention of public opinion and, therefore, expects to be sanctioned for failing to support the partner. So "fair play" will be expected only when the coalition totals more than 50 percent of the vote in the first round. Let us examine this conjecture with respect to the Communists. Table 7.4 replicates the analysis of Table 7.3 only for the coalition that comes in first in the first round. The R^2 of the model jumps from .09 to .58, and the coefficients are highly significant with the correct sign.

Were the Communists excellent tacticians after all? Did they behave as they should have whenever they were visible, that is, when they were the second party while the Left was about to win a seat? Based on the following pieces of evidence, it seems they did. An analysis of vote transfers reported in a special edition of *Le Monde* reveals two different patterns of vote transfer within the Left; it shows that Communist votes were transferred to the Socialist candidate, but not vice versa. Similar results were reported in survey findings by Jaffré (1980). At a more theoretical level, Bartolini (1984) discusses the positional advantage of the Socialists, who are located more to the center of the political arena and therefore collect all the Communist votes in the second round.

However, as Table 7.4 indicates, supporters of the other parties adopted exactly the same strategy as the Communists. In fact, the

TABLE 7.4. Cohesion of French coalitions winning in the first round as a function of different variables.

Coalition	Repr	N	R²	Cons	Victory	Prox	Adv.	Others
Left	PCF	98	.74	-.34 (-12)	.47 (14)	-.13 (-6.2)	.003 (1.2)	-.25 (-5.4)
Left	PS	109	.58	-.26 (-8.2)	.31 (11)	-.04 (-1.8)	.00 (.17)	-.24 (-4.5)
Right	UDF	118	.61	-.04 (-.7)	.19 (3.4)	-.13 (-7.7)	-.00 (-.82)	-.29 (-6.2)
Right	RPR	123	.70	-.13 (-2.7)	.28 (5.6)	-.12 (-8.2)	-.00 (-1.0)	-.35 (-6.9)

TABLE 7.5. Cohesion of French coalitions losing in the first round as a function of different variables.

Coalition	Repr	N	R²	Cons	Victory	Prox	Adv.	Others
Left	PCF	43	.22	.22 (1.7)	-.16 (-1.5)	-.06 (-.74)	-.00 (-.41)	-.46 (-2.3)
Left	PS	154	.08	.05 (1.4)	-.04 (-1.4)	.01 (.58)	.00 (.42)	-.16 (-2.8)
Right	UDF	87	.24	.11 (1.2)	.01 (.19)	-.08 (-2.3)	-.03 (-3.8)	-.29 (-2.5)
Right	RPR	120	.46	.23 (5.9)	-.13 (-3.8)	-.07 (-4.5)	-.01 (-5.2)	-.24 (-5.2)

fit of the model and the significance of the coefficients increase substantially when in each case the only constituencies considered are the ones in which each coalition was ahead in the first round. So for all parties, cohesion increases when it is needed, *when politics is visible*, that is, when a coalition is about to win the seat.

What happens when a coalition appears to lose in the first round, that is, when it receives less than 50 percent of the vote? Table 7.5 addresses this question. The fit of the model drops sharply, and the significance of the coefficients decreases. However, the competitive aspect of the interaction between coalition partners remains: the closer they are to each other, the more votes are missing in the second round. But the closer a coalition is to victory, the more partners undermine each other. Thanks to the theory of nested games developed in this book, interpreting this result is straightforward: each party attributes negative utility to its partner winning an additional seat. So whenever there is a viable formal excuse or whenever public attention is not concentrated on its behavior, each party undermines its own partner.

Thus, the differences in behavior according to whether one is ahead or behind in the first round can be attributed to different degrees of visibility that characterize the two positions. The pattern of helping one's partner when needed if politics is visible but undermining one's partner when it is invisible is reflected in the behavior of the UDF toward the Gaullists and of both partners of the Left, but it is not observable in the behavior of the Gaullists. These results indicate that the confusion in the political line of the leadership of the Communist party did not produce outcomes different from other parties. The strategies and behavior are fundamentally the same. Only the degree of precision varies.

A similar analysis could be made for the coefficients in Table 7.5. It should be kept in mind, however, that these coefficients are not statistically significant, and the results are therefore less reliable. Additionally, in this case, because the coefficients are all negative, there is no trade-off effect; therefore, the comparison is not interesting.

The previous analysis indicates that the political game is played on completely different terms according to whether each coalition partner's behavior is visible. This finding is in congruence with several other accounts. Sartori (1976), for example, argues that

Duverger's law(s) do not operate at the party system level, but only at the party level, because party strategies are part of visible politics; therefore, electoral considerations cannot be the exclusive basis for party choices. Inside a party, however, factions operate without any constraint (invisible politics); therefore, electoral considerations determine faction politics. Similar findings were reported in Chapter 6, where differences in visibility changed the nature of the game between the different Belgian political elites.

On the theoretical level, the principal-agent literature in economics builds on this distinction and on the opportunities that a loose monitoring procedure provides an agent (Jensen and Meckling 1976; Klein, Crawford, and Alchian 1978). This literature suggests that whenever a monitoring mechanism is installed, people's behavior will be modified to the extent that it is effective. No matter how self-evident this proposition seems, we have few empirical examples. The reason lies with the secrecy of invisible politics.

The fact that pure competition prevails in invisible politics has one very important consequence. The conventional wisdom is that French voters select in different ways in the first and the second rounds: "in the first ballot choose; in the second eliminate" (Converse and Pierce 1986, 356). However, my analysis indicates that the dominant party of the coalition that is ahead in the first round can expect to receive the support of its partner, whereas the dominant party of the losing coalition will have some of its partner's votes missing. The outcome will be that the winner of the first round can almost be assured of success in the second round. So there is no essential difference in outcomes between the first and second rounds. Indeed, out of the 448 constituencies, there were only 35 cases in which the Left came first in the first round but lost in the second (7.8 percent) and 18 cases in which the same happened to the Right (4.0 percent). In only 11.8 percent of the constituencies did the winner of the first round not win the seat.

One final observation can be made from a comparison of Tables 7.4 and 7.5: competitive behavior inside the Right is reduced where there is a Communist opponent, regardless of whether politics is visible or invisible. This finding is consistent with the fact that the Communist party has been considered outside the political system for most of the postwar period.

To sum up, the differential distribution of electoral strength at the local level accounts in large part for the variance of vote transfers inside coalitions. Parties (all parties) are more cooperative with their partner when a seat is at stake and more competitive when they have approximately equal strength. However, if the first round suggests that a defeat is probable (invisible politics), each party undermines its partner.

V. Nested Games and Alternative Explanations

Because French elections have been the subject of numerous studies, I can compare the conclusions of the nested games approach with those generated by other theoretical or empirical perspectives. The comparison can be made along different dimensions: empirical fit (whenever the data are sufficiently close to permit comparisons), parsimony, and congruence with existing theories.

In a survey reported by Jaffré (1980), the Communists appear to have voted strongly for the Socialists in the second round, whereas the Socialists did not reciprocate. Jaffré's findings are corroborated by data presented in the special edition of *Le Monde*, as noted earlier. Spatial voting explanations can account for such a difference: Communist voters have no choice but to vote for the Socialist candidate in the second round, whereas Socialists can choose the Right instead of the Communists. According to spatial voting theories and to Jaffré's empirical findings, cohesion in the Left is higher when the Socialists represent the Left. Jaffré's findings are presented on top of Figure 7.4A. The shaded areas indicate higher expected cohesion, and the visual comparison indicates the extent of the difference in the conclusions of these theories. As Figure 7.4A indicates, Jaffré finds that outcomes in the left part of the triangle (when the Left is represented by a Socialist) produce higher cohesion. However, he can neither explain variations in cohesion when the Left is represented by a Communist nor when it is represented by a Socialist.

Rochon and Pierce (1985, 439) examine the cohesion of French coalitions and conclude: "The general rule for both sympathy and cooperative behavior between the two parties will be that the coalition is most harmonious when it is least needed, that is when

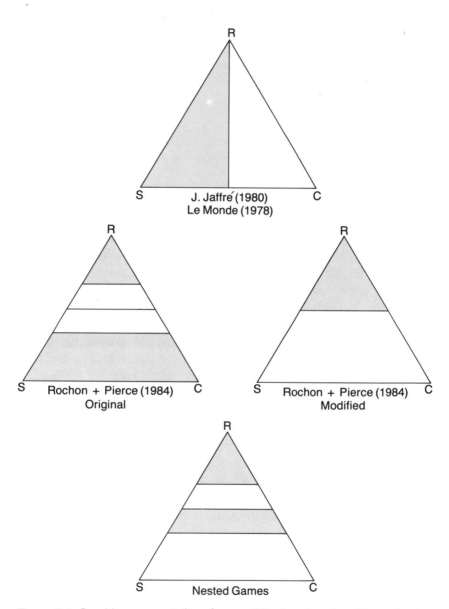

Figure 7.4 Graphic representation of competitive theories of coalition cohesion in a two-dimensional simplex.

the success of one of the two parties in capturing a legislative seat is not at stake." The data that Rochon and Pierce use are different from those presented in this chapter, and their analysis captures over-time rather than cross-sectional characteristics. Indeed, their data cover both the 1967 and the 1978 elections. To the extent that they generalize their results, however, some comparisons are possible.

Rochon and Pierce's argument is an equilibrium argument because they examine a long period of time (over ten years), and they assume that there are no temporal differences in party behavior.[14] I have argued elsewhere that the statement "the coalition is harmonious when it is not needed" cannot be true at equilibrium: the two partners of the Left would not enter into a coalition if it were to fail when needed (Tsebelis 1988a). If, by mistake, they did this once, their mistake should have been corrected in subsequent elections. The explanation that Rochon and Pierce (1985, 447–48) present for their empirical findings is social-psychological: "Party interests are defended in much the same way that individuals defend their own self-image when confronted with the unpleasant realization that a friend is outperforming them. . . . [They] try to undermine the success of the friend." This explanation cannot be an equilibrium argument either: if party candidates behave emotionally (Rochon and Pierce use the word *jealousy* to describe the behavior of Socialist candidates), they will be replaced by more rational candidates, who will maximize the votes of the party or the coalition.

We can now compare the comparative statics method of rational choice with the social-psychological explanation offered by Rochon and Pierce. What is unique in the rational-choice approach is that it assumes that individual action is an optimal response to existing constraints and to other individuals' actions. Optimal behavior is, therefore, self-explanatory. The analyst does not need to explain why an individual did the best he could under given circumstances. On the contrary, what needs to be explained

14. In fact, they examine the behavior of Socialist candidates after the 1967 and 1978 elections and find no statistical difference between the two samples, so they pool their data. There is nothing in the article to indicate that the behavior of Socialist candidates would or might have been different in between these two points in time.

is why people make choices other than optimal ones or why they have noninstrumental motives like jealousy when the political stakes are so high. As Figure 7.4B indicates, the original formulation by Rochon and Pierce expects higher cohesion in the upper and lower parts of the triangle when a seat is not at stake.

Because the statement that "the coalition is most harmonious when it is least needed" cannot be true if actors are rational, I have proposed a minimal verbal modification that is consistent with Rochon and Pierce's empirical findings: "The coalition is most harmonious when it is about to lose a seat" (Tsebelis 1988a, 236). As Figure 7.4C indicates, my reformulation of the Rochon-Pierce findings expects higher cohesion in the upper half of the triangle, when seats are about to be lost by the Left.

Figure 7.4D presents a simplified version of the nested games approach, in which cohesion is expected to be higher when the Left approaches the fifty-fifty split from above or when it is far behind. Why are these figures and conclusions so divergent?

First, let us concentrate on Jaffré's findings. As we saw in Section IV, there is no fundamental difference between the behavior of Communists and that of other parties. The same equation can account for the behavior of all parties. What changes is the size of coefficients, not the sign. Both Jaffré and *Le Monde* find Communist behavior different because they are interested in the description and not in the explanation of vote transfers. Therefore, the appropriate explanatory variables are absent from their analysis.

Spatial explanations have been offered to account for the difference in the pattern of vote transfers inside the Left (Bartolini 1984; Rosenthal and Sen 1973, 1977). Indeed, we saw that overall the Communists are more faithful partners than the Socialists. However, policy explanations are not sufficient to explain variations in party behavior. As we saw, all parties transfer their votes better when their partner needs them and politics is visible and worse when they get equal or almost equal vote shares with their coalition partner. Spatial voting explanations cannot explain why the same party sometimes transfers its votes effectively and other times does not. The theory of nested games provides the additional reason for defective vote transfers: intracoalition competition, which is the result of the closeness in scores of the two parties in the first round.

TABLE 7.6. *Proximity of the two partners of the Left when they approach victory.*

	Socialist lead	Communist lead
General	.908	.916
(number of cases)	(302)	(146)
victory > .95	.909	.938
(number of cases)	(158)	(85)
victory > .97	.907	.943
(number of cases)	(113)	(53)
victory > .99	.910	.959
(number of cases)	(38)	(11)

Note: Victory = 1 means a 50–50 split of the vote between coalitions.
Proximity = 1 means equal split of the vote between partners.

The conventional wisdom, expressed by Jaffré and *Le Monde*—that the Communists transferred their votes more effectively than the Socialists—can be explained to a certain extent, but can also be challenged. If the Socialists appear to be more competitive than the Communists, it follows that when the Left is led by a Communist at the constituency level, the Socialist is usually slightly behind, whereas when the Socialist is ahead, the Communist is significantly behind.

Table 7.6 confirms this expectation. The first column indicates that in constituencies in which the Socialists lead the Left, the Communists remain some ten percentage points behind, regardless of how close the coalition is to victory. The second column demonstrates that in constituencies in which the Communists lead, the closer the coalition is to victory, the smaller the difference is between the two parties in the first round. So the reason that the Communists transfer their votes better than the Socialists is the distribution of votes in different constituencies: in Communist-led constituencies, Communists and Socialists receive approximately the same number of votes; in Socialist-led constituencies, the Socialists are well ahead of the Communists.

The Left was close to victory in 1978 because of the rapid growth of the Socialist party. This, however, had negative effects on the cohesion of the Left. Political commentators at the time

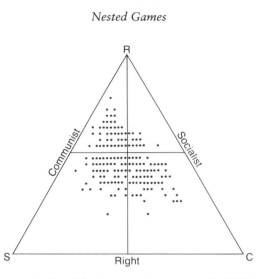

Figure 7.5 Representation of first round electoral results (PS, PCF, Right) in a two-dimensional simplex.

stressed the fact that it created reactions on the part of the Communist leadership. This analysis shows that the Socialists' rapid growth created an additional problem for the unity of the Left: the Socialists started making claims over constituencies traditionally represented by Communist candidates. This created tensions and resulted in defective vote transfers from the Socialist party.

Figure 7.5 represents graphically the first round electoral results in the two-dimensional simplex (triangle) of Section I. The figure focuses on the relative strength of the two left-wing parties (PCF, PS).

Figure 7.6 presents the same first round electoral results, but the focus this time is on the relative strength of the parties of the Right. The different constituencies are plotted on a triangle, which represents the UDF, the RPR, and the Left. The scale of Figure 7.6 is the same as that in Figure 7.5 to facilitate visual comparisons. Note the difference in the spatial distribution of the two coalitions when comparing the two triangles. The Right is expanding along and below the horizontal axis (visible politics); the Left is concentrated around the origin and expands along the vertical axis. But as stated earlier, whenever the distribution of election results runs along and below the horizontal axis, the coalition has maximum

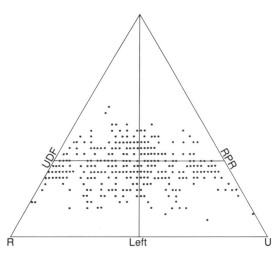

Figure 7.6 Representation of first round electoral results (RPR, UDF, Left) in a two-dimensional simplex.

cohesion. Whenever the distribution runs along the vertical axis, competition increases.

So the victory of the Right in 1978 can be attributed to two factors: an overall better quality of transfers and a more favorable (asymmetric) distribution of strength between the partners, which further improved the quality of transfers.

The explanation of the discrepancies between Rochon and Pierce's findings and my own is more difficult. There are several possible lines of argument. First, their data differ from mine; therefore, the two theories account for different phenomena, or in the epistemological jargon, they are incommensurable (Kuhn 1962). Second, the two theories discuss the same phenomena and leave the decision on which approach is best to the usual empirical criteria of goodness of fit.[15] Third, one could combine the empirical findings from the two theories into a more general theory of electoral cycles (Tsebelis 1988a). Arguments can be presented for and against each of these three lines of reasoning.

15. Although any comparison according to the standard criteria of goodness of fit (R^2) and statistical significance (t) would be extremely advantageous for the nested games approach, it would be unfair to Rochon and Pierce because they have fewer data points and deal with individual cases rather than aggregate data.

In a voluminous analysis of French elections, Converse and Pierce (1986) dedicate two chapters to the study of electoral participation and the flow of the vote in the second round, comparing their findings with earlier analyses by Rosenthal and Sen (1973, 1977). Rosenthal and Sen's conclusion is that spatial voting models explain participation in the second round, whereas Converse and Pierce (1986, 351) find that participation in the second round "can be accounted for in terms of partisan sympathies but not by sense of left-right distance." Converse and Pierce (1986, 353) explain the difference in conclusions by noting that Rosenthal and Sen use aggregate data and "clump" left-right locations of parties.

Both Rosenthal and Sen and Converse and Pierce investigate different election years than the present chapter.[16] Moreover, both use either individual or aggregate data for the right-left positions of voters and parties, also absent from the present chapter. Finally, the dependent variables in their studies (abstentions, spoiled ballots, or votes for parties in the second round) are connected with but not identical to cohesion, the dependent variable in this study. Any comment on their findings, therefore, should be considered tentative. Because the final conclusion of Converse and Pierce is that there is room for further research, some of the conclusions of the present investigation may be relevant.

Converse and Pierce demonstrate conclusively that spatial voting models alone cannot explain abstentions and flows of the vote. In addition, the best single model in terms of fit was Rosenthal and Sen's (1973) "heuristic" model, in which the closeness of a race (measured by the difference between front-runner and runner-up) and choice (measured by the number of candidates in the second round) were the independent variables. Some additional explanatory variables are needed. It seems that this is exactly what Converse and Pierce capture with a "partisan sympathy" variable.

Partisan sympathy, however, has the status of an independent variable in a social-psychological study in which it can be measured by the appropriate survey questions; in a rational-choice

16. However, one of the elections covered by these studies was the 1967 election, which, in terms of its general political climate, was similar to 1978 because most constituencies were disputed between a candidate of the Left (usually Communist) and a candidate of the Right (usually Gaullist).

account, feelings of sympathy or competition with other parties and, more to the point, the behavioral consequences of such feelings must be *derived* from the features of the political environment and the rationality of the actors. This is precisely what the proximity variable in my model captures: feelings of sympathy vary inversely with the competition between partners during the electoral campaign, and this competition is a function of the closeness between partners. So the nested games approach pushes Converse and Pierce's findings one step further: it provides an explanation for the feelings of sympathy or competition that Converse and Pierce take for granted.

The model in this chapter presents more parsimonious and empirically accurate results than previous approaches. Moreover, it presents an account for partisan sympathy that is based on contextual forces operating at the constituency level: the distribution of votes between parties and between coalitions.

VI. Electoral Laws as Institutional Design

So far I have used the first kind of nested games (games in multiple arenas) to describe coalition cohesion as the optimal response of parties to each other and to the electoral system. In this final section, I reverse the argument and concentrate on the second kind of nested games: institutional design. I demonstrate that because the electoral system consistently produced results unfavorable to the Left, the Socialist government introduced proportional representation (PR) in 1985; for the same reason, the right-wing coalition reestablished the two-round majority electoral system (TRMES) when it came to power one year later. According to the typology presented in Chapter 4, this section argues that the electoral system was a consolidating institution: once in power, each coalition tried to consolidate its own position using the electoral system as the principal instrument.

The Socialist party emerged from the 1978 elections as the strongest single party in France; its governmental aspirations, however, were seriously handicapped by the TRMES. The ally of the Socialists was the weakest of the remaining three parties. Thus, the outcome of the alliance was the defeat of the Left. Simulations of the electoral results indicate that if the 1978 elections had been conducted under proportional representation, the Left would have won (Meyer 1978).

The potential victory of the Left in 1978 under PR is not a sufficient reason for the establishment of the PR electoral system by the Left in 1985. Because the performance of institutions is evaluated over the long run, as Chapter 4 argues, one has to show that the TRMES worked against the long-term interests of the Left in order to make a convincing case. Further simulations of electoral results indicate that under PR, the Left would not have won in 1981 (Bon 1985). Therefore, it would have been unwise to change the electoral system for partisan reasons given that change does not seem to produce results that are systematically biased one way or the other. Moreover, it can be argued that the change to PR occurred because the Left had promised it both in the common program of government and in the electoral campaign of François Mitterrand.

Simulations of electoral results under different electoral systems have produced the following results: generally, the TRMES compared to PR produces underrepresentation of the defeated coalition, underrepresentation of the forces that refuse political polarization,[17] underrepresentation of the marginal forces of the Right and the Left, and underrepresentation of the smaller partner inside each coalition (Parodi 1983). Such simulations, however, express only an indicative value because they assume that people vote the same way regardless of electoral system: there is no strategic voting.

In order to understand the strategies that led to the two successive rule modifications, we must examine more closely some of the political events of the 1980s. In 1981, Mitterrand for the third time contested the second round of a presidential election and, for the first time, won. His election marked the first time that the Left occupied the office of the president under the Fifth Republic. He immediately dissolved Parliament (elected in 1978, it had a hostile majority) and led the Left to its first legislative victory under the Fifth Republic. The victory was magnified by the TRMES to the

17. Such forces were the center parties, which had been almost eliminated since 1974, and the presidential race between Giscard and Mitterrand.

point that the Socialist party (with 282 seats) controlled the absolute majority in the National Assembly without depending on the 43 Communist votes.

Although Communist support was not formally needed, Mitterrand invited the Communists to participate in government. The coalition lasted until the summer of 1984, when the Mauroy government was replaced by the Fabius government, in which there were no Communist ministers. During that time, the tensions between the two parties increased, and consequently, the Communists started hinting that republican discipline (voting for the front-runner of the Left in the second round) was not assured. Duverger (1982) had long ago advised the Socialists to change the electoral system so that they would not be hostages of the Communists.

At this point, I note the following:

Mitterrand and the Socialist party. They could use the magnifying effect of the front-runner inside each coalition to obtain an important share of the vote (assuming Communist support). However, all the polls indicated there was no chance that the Left (an alliance of Socialists and Communists) could win the 1986 elections. Moreover, the Communist party was in serious decline.[18] Even if its support were assured, the balance of forces under TRMES would have been extremely favorable for the Right. *The Economist* (November 3, 1984) projected the European election results for national elections under different electoral systems, demonstrating that modification of the electoral system was the dominant strategy of the Socialists: with TRMES, the Socialists would receive 45 seats, the Communists 26, and the Conservatives 403. With PR, at the national level, the corresponding numbers would be 100, 53, and 205. Nevertheless, the more important reasons for the change in the electoral system were of a strategic, not arithmetic, character: the political and economic situation had provoked serious disagreements and divided the Left. The Socialist party had greater mobility and coalition potential because of its pivotal position along the right-left axis. The TRMES polarized the various political families, preventing the Socialist party from

18. In the 1984 European elections under PR, the Communist party received 11 percent of the vote (Knapp 1987).

exploiting its pivotal advantages. For example, if the two parties of the Right were not able to obtain a majority of the seats, they would be forced to choose between a coalition with the Poujadist party of Jean Marie Le Pen (which was a very uncomfortable position) or a coalition toward the center.[19] This possibility would increase the Socialists' chances of remaining in government, increasing the political power of the president of the republic, who would then have the power to build and maintain the coalition in government.[20]

Communist party. The party found itself more and more isolated and marginalized. The choice was not whether to participate in any government (one cannot be sure whether it wanted to or if it could find any allies), but how it could obtain more seats in the National Assembly. The characteristics of TRMES and PR made such a choice clear: the smaller party in the smaller coalition is practically wiped out by TRMES. Consequently, the Communist preference was for PR.

RPR and UDF. The parties of the Right held the opposite interests from those of the Socialist party. They knew that TRMES would magnify their victory, assuring them a comfortable majority. They also knew that TRMES would increase their cohesion, making them independent of any support from the Right (National Front of Le Pen) or the Left (Socialist party). Therefore, the choice was clear: both parties were for TRMES, with the Gaullists more favorably disposed because they would receive the lion's share from the alliance.

These were the preferences of the different parties. The two parties of the Left, for different reasons, preferred PR; the two parties of the Right preferred the TRMES. The official rhetoric from both sides used arguments concerning the function of institutions, fairness, and government stability. However, because the Left had an overwhelming majority in the National Assembly, the

---

19. The only Socialist leader who did not share the preference for a PR system was Michel Rocard, who at the time resigned from the post of minister of state (Ministre d' État) and agriculture, explaining that under PR, the Socialists would never regain their majority. At the time, there were questions in the press concerning the real reasons for his resignation.

20. The opinion of scholars regarding the power of the presidency is divided. Some argue that the president would have lost power because he would not be supported by a stable majority. See Duverger in *Le Monde* (May 29, 1985).

dispute was easily decided: the new electoral law was passed July 11, 1985, less than a year before the upcoming legislative elections. It mandated PR and an increase in the number of seats in the National Assembly from 491 to 577.

Is this account fair to the parties involved, or is it a cynical misrepresentation of their motives? What were the "real" motives of the actors? This question is difficult to answer directly because no actor would admit partisan motivations. However, indirect evidence can be presented. Presumably, if the choice over electoral laws involved principles, different actors' opinions should remain stable over time; if, on the contrary, it involved tactical considerations, as previously enumerated, the political actors' opinions should vary.

Because Mitterrand was one of the principal actors in the electoral reform, I concentrate on his expressed opinions and propositions. The first piece of evidence comes from Mitterrand's speech to the fourth national conference of the UDSR in 1950. It lays out the rules of choice of electoral laws: "I do not include any doctrinal element. The electoral scheme that I choose must result from a political opinion. . . . In fact, there are certain points which make this option necessary. First, what is the interest of the Nation? Then, what is the interest of the majority to which I belong? Finally, what is the interest of my party? And I will make up my mind when I have answered these three questions" (Chagnollaud 1985, 95, translation mine).

It seems that the answers to these three questions varied over time because Mitterrand frequently changed his mind about the electoral system. In the second UDSR conference (1948), he called PR "néfaste" (nefarious) and declared his support for a TRMES. As a minister in 1950, he refused to sign an essentially majoritarian electoral law with some elements of proportional representation. Later in 1950, he proposed a one-round majority system at the *départemental* level (multimembered constituency), which was rejected by his party. In 1951, he voted for the "apparentements" electoral system.[21] Between 1953 and 1955, he undertook several

---

21. The "apparentements" system was a mixture of proportional and majoritarian elements that permitted the merging of different parties in different areas. The purpose and the outcome of this system were to exclude the Gaullists and the Communists from the majority of the seats, creating a centrist majority. See Chapsal and Lancelot (1969).

initiatives to promote a majoritarian system at the constituency level. In 1958, he supported the Gaullist reform of the electoral system. In 1968, he made his first timid statements in favor of PR. In 1970 and 1971, his statements became very clear, and his preference for PR has remained stable ever since.[22]

This account indicates that the electoral system did not have the status or the stability of the constitution in the minds of political actors.[23] Indeed, the period of stability of the TRMES under the Fifth Republic (twenty-seven years) was exceptional in French history. Only once did the electoral system last longer (1889–1919). By coincidence, it was the same TRMES adopted by the Fifth Republic.[24]

Mitterrand's electoral reform, however, was neither extremely successful nor long lasting. It was not successful because the parties of the Right managed to get a slight majority of seats (291 out of 577) in the new National Assembly. It was not long lasting because the new majority restored the TRMES one year after its abolition (July 10, 1986).

The motivations behind the new modification are obvious from the previous account: it was a conscious effort by the Right to consolidate its position as the majority. What merits a more detailed description are the political maneuvers that made this modification possible. The following demonstrates that the new prime minister and leader of the majority, Jacques Chirac, masterfully utilized existing institutions to achieve his goal.

The right-wing government coalition received only 3 seats over the required majority of 288 in the 1986 elections. But this majority was not solidly behind the prime minister's plan to restore the TRMES. First, the allies (the independents in particular) were not as enthusiastic as the Gaullists over the TRMES. Second, there was discussion about reducing the number of seats to 491 instead of 577, which would remove some members of the National Assembly from reelection. Third, and most important, the need for redis-

22. For the full account and its sources, see Chagnollaud (1985, 101–3).
23. Chagnollaud gives similar historical accounts of other presidential opinions during the French Fifth Republic.
24. The electoral system has been modified twelve times in France since 1871, seven times in this century: an average of one modification every eight or twelve years, depending on the basis of the calculations. See LeGall (1985).

tricting was becoming obvious, and the allies were afraid that Charles Pasqua, the new minister of the interior and faithful friend of Chirac, would base his calculations heavily on the interests of the Gaullist party.

For Chirac, the situation was as follows: the opposition, both from the Left (Socialists and Communists) and from the Right (National Front), was united against his electoral reform; and the current members of his majority in the National Assembly who would not be reelected under TRMES would most likely vote down the proposition of electoral reform. Even in the unlikely case that most would demonstrate an exceptional spirit of self-sacrifice, three defections were enough to defeat Chirac's plan. For these reasons, Chirac knew that an open debate of the electoral reform on the floor of the National Assembly would substantially modify his project, relinquishing significant political advantages to the opposition.

The only way out of this impasse was not to allow discussion of the electoral reform before the National Assembly. In Chapter 6, I discussed the advantages held by monopolists who can present their own propositions and have them voted up or down without discussion. In that particular kind of situation, any proposition remotely superior to the status quo can be adopted. In the case of Chirac's reform, however, there was practically no majoritarian electoral reform that could have been adopted by the majority of the National Assembly elected in 1986.

Therefore, in addition to not accepting any amendments (closed rule), Chirac took advantage of an article in the constitution and resorted to blackmail. Article 49.3 mandates that any *project* of law may be transformed by the government into a question of confidence.[25] According to article 49.3, the governmental project is accepted without a vote. If a motion of nonconfidence is presented by the floor, this motion (not the governmental project) is debated and voted. The role of article 49.3 is to change the subject of discussion and force the deputies of the majority to support the government despite their disagreement on a specific piece of legislation.

25. If a bill is introduced on the floor of the National Assembly by the government, it is called a project; if it is introduced by a deputy, it is called a proposition.

Figure 4.2 can be used to visualize the situation. Consider that player 1 represents Chirac's solid coalition (the RPR and part of UDF), and player 2 represents those of his allies who oppose the electoral reform. Introducing the electoral bill shifts the situation from Figure 4.2A to 4.2C, where some of the deputies in Chirac's fragile majority would have voted with the Left (player 3) to defeat the bill. Article 49.3 forces the battle back onto the original axis of Figure 4.2A; it mandates that the issue under consideration is not the bill, but the government itself, on which players 1 and 2 are in coalition. To use the terminology of Chapter 4, the electoral reform was a consolidating institution.

Even this method of adoption, however, would not have been successful with the deputies who knew with certainty that they were doomed in the next election. The government's project had to hide its Achilles' heel: the redistricting operation. Providing information about redistricting would have permitted deputies of the majority to make the appropriate calculations and would have led to loss of votes. The government's second step was not only to postpone the redistricting—the problem of insufficient votes would appear in any subsequent attempt to pass the project—but to remove it from the jurisdiction of Parliament. Article 38 offers constitutional means to perform this operation because it provides the government with the authorization to legislate by decree in a certain area, in this case, redistricting.

In Chapter 4, we saw the role of uncertainty in the adoption of institutions: when uncertainty increases, efficient institutions become more likely. The redistributive effects of the electoral system between the Right and the Left were clear, which is why both actors took the initiative of reform whenever the opportunity was available. However, although the reform was profitable for the Right overall, it was not for all its representatives in Parliament because of the issue of redistricting. Let us ignore for a moment the Left and focus exclusively on the Right. From the point of view of the collective interests of the Right, the electoral reform was necessary because it promoted the interests of all its members. Speaking about the Right exclusively and using the terminology of Chapter 4, the TRMES was an efficient institution. In order to be promoted, uncertainty about which particular members of Parliament would be hurt by the reform was required. Article 38 provided the government with the required "veil of ignorance" and enabled it to promote the efficient (for the Right) institution.

Chirac's rhetoric included the argument that the TRMES was a "pillar" of the constitution. It was, however, Chirac's masterful use of institutions and the combination of articles 49.3 and 38 that allowed his victory. As expected, the Socialists put forward a motion of nonconfidence, which received only 284 votes and was defeated (Avril and Gicquel 1986, 175). The acceptance of the TRMES was at the same time a partisan triumph for the Gaullist party because through article 38, Charles Pasqua became the exclusive master of the redistricting process.

This account of institutional change confirms several points made in Chapter 4. First, institutions are the outcomes of conscious design. Second, they can be used as weapons in a political battle; their use becomes increasingly necessary when a majority is thin and problematic, as it was for the French Right after 1986. Third, political actors will create institutions to consolidate their positions or amplify their victories. Fourth, the role of information for the creation of different types of institutions (efficient as opposed to redistributive) and the conscious manipulation of information (article 38) on the part of some players (the government) is crucial in order to promote the desired institutions.

VII. Conclusions

In the beginning of 1978, opinion polls were uniformly indicating that the Left was ahead in a close race between the two coalitions. In March, the Left lost the election. The situation was aptly captured by the title of an important sociological analysis of the results: *France of the Left Votes to the Right* (Capdevielle et al. 1981). Why did it happen? The account I present indicates that the rapid growth of the Socialist party, which was expected to be the driving force in the electoral victory of the Left, weakened the leftward swing because it created in the first round a distribution of votes that was unfavorable for the cohesion of the Left and generated bad vote transfers.

Mitterrand modified the electoral system that had produced

such unfavorable results and replaced it with proportional representation. The Right's 1986 electoral victory reversed the situation, and the previous electoral system was reinstalled.

The theory of nested games thus provides not only an accurate description but also an explanation of how the French electoral system works and how political parties all behave the same way. Cohesion of coalitions (of each coalition) increases when it is needed (when coalitions have equal vote shares) and politics is visible. Competition between parties (all four political families) increases when the partners inside each coalition are of equal strength.

The theory can explain variations in the cohesion of coalitions that spatial voting theories cannot. Indeed, according to the latter, there is no reason for such variation. Moreover, the nested games approach is capable of demonstrating why some social-psychological explanations relying on concepts like jealousy are wrong and what lies behind independent variables like feelings of sympathy or competition between parties: the distribution of votes between partners and between coalitions. Finally, the nested games approach promises further applications, explored in Chapter 8.

Appendix to Chapter 7: Definition of Variables

rpr: number of votes for the Gaullists (in the first round)

udf: number of votes for the Giscardians (in the first round)

ps: number of votes for the Socialists (in the first round)

pc: number of votes for the Communists (in the first round)

reg: number of registered voters

fround: number of voters in the first round

sround: number of voters in the second round

tleft: number of votes for Socialists, Communists, and allies in the first round

tright: number of votes for Gaullists, Giscardians, and allies in the first round

left: number of votes for the candidate of the Left in the second round

right: number of votes for the candidate of the Right in the second round

victory: $1 - \text{abs}^1(\text{fround}/2 - \text{tleft})/\text{reg}$

proxl: $1 - \text{abs}(\text{ps} - \text{pc})/\text{reg}$

proxr: $1 - \text{abs}(\text{udf} - \text{rpr})/\text{reg}$

otherl: $(\text{tleft} - \text{ps} - \text{pc})/\text{reg}$

otherr: $(\text{tright} - \text{udf} - \text{rpr})/\text{reg}$

1. abs = absolute value.

advl: dummy variable with value 0 if the adversary is Socialist

advr: dummy variable with value 0 if the adversary is Gaullist

cohl: (left − tleft)/reg

cohr: (right − tright)/reg

Chapter Eight

Conclusions

Instead of summarizing the major findings of and suggesting new applications for each chapter of this book, I pull together from each chapter the recurring themes. This book offers a rational-choice approach to issues of comparative politics, with particular reference to Western European politics. Substantively, it deals with two distinct themes: political context and political institutions. Consequently, I follow this outline here: Section I discusses the issue of rationality, Section II deals with questions of political context (that is, games in multiple arenas), Section III concerns political institutions (that is, institutional design), and Section IV offers a final remark.

I. Rationality

This book started with the assumption that people are rational, that is, that they are goal oriented and choose the optimal means to achieve their goals. The implicit assumption of rationality is the common denominator of most of social science research. The book departed from the bulk of comparative politics research by making the rationality assumption explicit, drawing out all its consequences, and even using the most demanding of them to derive testable propositions about different countries, parties, and institutions.

The rationality assumption and its consequences (essentially, game theory and comparative statics) already have had and will continue to have an important impact in political science. There are two reasons for this impact. The first is that rationality is not a latecomer in politics, the social sciences, or philosophy. One can trace this assumption back to the framers of the American constitution, the philosophers of the Enlightenment, or ancient Greece; indeed, almost all of Western civilization is founded on it. Rational choice simply pushes the idea of rationality to its logical limits. In this sense, it is part of a long-standing tradition and is close to other mainstream approaches in political science.

The second and more important reason for the impact of the rationality assumption is that rationality, together with its consequences (in other words, rational choice theory), can highlight serious omissions in or mistakes of intuitive reasoning. The paradoxes pointed out by rational choice have over the years proliferated, indicating that our intuitions are not reliable guides to understanding the world around us. For example, such familiar phenomena as social groups became the object of investigation rather than being taken for granted after Olson's (1965) work on collective action. The incompatibility of certain elementary characteristics of rationality and fairness discovered by Arrow (1951) shed a new light on democratic and other institutions. Voting, abstention, rational ignorance, and information are now investigated in new ways. The common denominator to all these new approaches is the belief that intuitive reasoning alone is not sufficient, that a more rigorous deductive reasoning is desirable whenever possible.

Using a historical perspective, one can compare the impact of major research programs like behavioralism and rational choice. The lasting impact of the behavioral revolution was that it changed our conception of what constitutes adequate empirical evidence. Instead of historical examples that corroborate a general statement, today it is widely understood that a whole population of events or a random sample from it has to be studied in order to come to empirical conclusions. The lasting impact of the rational-choice research program is that it changes our conception of what constitutes theoretical and coherent reasoning. It is not true that theoretical and coherent reasoning can be found only in formal models, just as it is not true that sound empirical conclusions can be reached only through sophisticated statistical techniques. What is true, however, is that both the methodological sophistication of the behavioralist program and the deductive rigor of formal models can help contemporary researchers reach conclusions that only the most powerful minds reach without such tools.

I believe that the rationality assumption will have an enormous organizing impact on the social sciences. Today the rational-choice agenda is criticized as being either trivial or too demanding and restrictive. This book offers two reasons why such accusations are misguided. The first was presented in Chapter 2, where I argued that processes such as learning, natural selection, heterogeneity of individuals, and statistical averaging can lead to the same outcomes as rationality. Whenever such conditions hold, rationality offers a good approximation of reality. The second reason was offered in the empirical chapters of the book (5, 6, 7), which concretely demonstrated the explanatory power of the rationality assumption.

In every chapter, a puzzle arose because I assumed the actors to be rational; yet their observed behavior seemed nonoptimal. Therefore, the rationality assumption essentially defines the agenda of this book.

Chapter 5 inquired into the seemingly suicidal political behavior of Labour party activists. This behavior requires explanation only because I have explicitly adopted the assumption of rationality. Similarly, if the activists' behavior seemed strange, it is because one normally assumes that their behavior should make sense. If we were able to assume that activists were "fanatics," as B. Webb describes them, or that their behavior was not instrumental because it was determined by principles, as Epstein (1960, 385) claims, there would be no puzzle.

Chapter 6 focused on the accommodating behavior of Belgian elites. Within the consociational literature, Belgian elites have been assumed to be independent of the masses they represent and to possess the capacity of compromise and accommodation. If this description were correct, we would expect Belgian elites never to initiate political conflict. But the empirical literature indicates that sometimes they do. Another explanation of elite accommodating behavior in Belgium was cross-issue bargaining. If this explanation were correct, elites would not need to engage in institution building. Institution building would have been a redundant and, because it consumes resources, a suboptimal activity.

Chapter 7 investigated the cohesion of French electoral coalitions. In the 1978 French election, the votes were not transferred between partners of coalitions in the way surveys indicated they should have been or spatial voting theories would have predicted. According to these theories, there should be no variation in the quality of vote transfers across constituencies. My own research, as well as the findings of Converse and Pierce (1986), indicate that such variations exist. Converse and Pierce used partisan identities to explain these variations. Although partisan identities are independent variables in their framework, in this book, I explained them as optimal behavior under the constraints of the electoral system and the distribution of votes at the constituency level.

The rationality principle and derivations from it provided the principal instruments enabling us to see why some empirical questions asked by previous research were meaningless. Indeed, we saw in Chapter 5 that there had been a substantial academic debate on the relative strength of the NEC versus different Labour party constituencies. The arguments in this debate involved observed frequencies of disagreement between constituencies and the NEC. Some academics and journalists employed time and resources to investigate "the secret garden of British politics." They used ingenious techniques to discover not only visible and public disagreements, but invisible and private ones, such as messages or phone calls from the NEC to the constituencies. However, the frequency of public disagreement does not indicate power; it indicates only informational imperfections in the game between the constituencies and the NEC. The observed frequency of disagreement between two actors is in fact irrelevant to any underlying distribution of power. In this case, the rationality assumption was used to evaluate the validity of empirical arguments.

The rationality principle and the corresponding comparative statics method can also lead to the discovery of mistaken theoretical arguments. In Chapter 7, I discussed one social-psychological explanation of the relation between allies within each coalition offered by Rochon and Pierce (1985). They explained tensions inside each coalition as an indication of the Socialist candidates' jealousy of their Communist allies. The conclusion of their account was that "the coalition is most harmonious when it is least needed" (Rochon and Pierce 1985, 439). I explained that this description cannot be true at equilibrium and cannot be held over a long period of time. At least one of the two allies would have rejected an alliance that offered nothing when it was needed. The rationality principle and comparative statics analysis thus enabled us to test the validity of other theoretical arguments.

Finally, because of the rationality principle, I decided against accepting easy historical or cultural explanations. In Chapter 7, I explained the anomaly of Communist behavior. A historical explanation of why the Communists were behaving differently than were other parties was possible: they had no strategy (up to the very last moment). Thus, it made sense that their behavior resembled random noise. Moreover, an entire array of arguments referring to the Communist subculture would have provided another possible account for Communist behavior. The explanation of Communist behavior offered in Chapter 7 is quite different. The Communists behave exactly the same way as the other parties when the model is correctly specified and the constraints imposed by public opinion are taken into account. This point demonstrates the interchangeability of actors mentioned in Chapter 2. Communist behavior is explained by rationality and constraints alone, without reference to any other attribute not explicitly included in the model. Communists, Socialists, Gaullists, and Giscardians behave exactly the same way (with variations in the size of coefficients), and their behavior is determined by the same factor: whether they seem to win or to lose in the first round.

The rationality assumption is therefore responsible for the selection of research topics, tests of theoretical or empirical arguments, and some of the conclusions in this book.

II. Political Context

I used elementary game theory in this book to model the interaction between different political actors. Game theory differs from other variations of the rational-choice program (decision theory, social choice theory, or economic theory) because it models explicitly the interaction between agents who can make contingent choices. Two of the games studied in Chapter 3 (prisoners' dilemma and deadlock) possessed dominant strategies: each player has an unconditionally best strategy available. For the other two (chicken and assurance), each player's optimal behavior depended on the opponent's choice.

Games are usually studied for their equilibria; no attention is paid to variations of payoffs. This book adopted a different perspective. It demonstrated that when correlated (or contingent) strategies are possible, payoff size determines the likelihood of different strategies being adopted. The same result can be proven in the case of iterated games, following Fudenberg and Maskin's (1986) proof of the folk theorem (that any individually rational outcome can be supported as a perfect equilibrium in an iterated game with a sufficiently long time horizon). Chapter 3 proved two fundamental propositions, which I then used throughout the book. These two propositions (3.6 and 3.7) relate the magnitude of payoffs to the likelihood of different strategies in iterated games, and they are necessary conditions for the development of games in multiple arenas.

Games in multiple arenas describe the situation in which one actor is simultaneously involved in two or more games or, equivalently, situations with externalities. The events or the actions of a third player in one arena influence the payoffs of the players in the principal arena, and the magnitude of the payoffs determines the players' strategies. I provide examples of the applicability of the theory of nested games, then recapitulate the major findings of the previous chapters.

Applicability. In Chapter 5, the game between MPs and activists was nested inside the electoral competition between Labour and Conservatives at the constituency level. The safeness of the seat determined the payoffs of the two players and, therefore, the likelihood of adopting different strategies. Similarly, the game between activists and the NEC was nested inside the national competitive game between the two major parties, and this competition determined the resources available to the NEC and the activists at each point in time.

Another way of thinking about Chapter 5 is that it presented comparative statics results, comparing safe and marginal constituencies or centralized British and decentralized American parties. Indeed, in each constituency, the relative strength of Labour (that is, an exogenous event) determines the different payoffs of the game, and these payoffs determine, in their turn, the behavior of the actors (activists and MPs). The diagrammatic exposition of the argument in Chapter 5 was presented in Figure 3.3A.

Similar approaches to political situations, without the nested games framework, can be found frequently. Such accounts present structural similarities with the analysis in Chapter 5.

In international relations, the theory of hegemonic stability claims that "hegemonic structures of power dominated by a single country are most conducive to the development of strong international regimes whose rules are relatively precise and well obeyed" and that "the decline of hegemonic structures of power can be expected to precede a decline in the strength of corresponding international economic regimes" (Keohane 1981, 132).

One explanation of the previous quotes is that the hegemon and the smaller countries enter into an iterated game.[1] A third player, nature or history, decides whether the hegemon is strong or weak.[2] In the case of a hegemonic power, smaller countries depend on the hegemon, behaving in a cooperative way that generates "strong international regimes" and "rules." Dependence means that cooperative or noncooperative behavior on the part of the hegemon is essential for the welfare of the small countries. In terms of the vocabulary of this book, for the small countries, $R \gg P$; therefore, according to propositions 3.6 and 3.7, small countries behave cooperatively. The decline of hegemonic power decreases the dependence of smaller countries and increases the likelihood of noncooperative strategies.

Another example can be offered by the dominant role of business in capitalist societies. Lindblom (1977) insists that business is not just another lobby and that public officials must pay particular attention to the political demands of firms. This idea is not new. Marxist analysis has stressed for a long time the fact that the "bourgeoisie" is the dominant class in capitalism. Przeworski (1980) has interpreted the Gramscian concept of hegemony along these lines, arguing that in capitalist societies, every other class

1. Most analysts claim it is a prisoners' dilemma game, but as Chapter 3 showed, because the game is iterated, what matters is the magnitude of the payoffs, not their order.

2. Just as a third player decided whether a constituency was safe or marginal in Chapter 5.

depends on those who own the means of production because capitalists alone have the power to invest.[3] In terms of games in multiple arenas, some initial distribution of wealth and means of production made other classes dependent on the bourgeoisie. For every social group or class in its prisoners' dilemma game with business, $R \gg P$, and according to the previous arguments, cooperation is likely.

Many authors explain the breakdown of social consensus in the 1970s as the result of an overvalued dollar leading to the end of fixed exchange rates, the sudden rise of the price of oil, growing protectionism, inflation, recession, depression, and economic crisis (Berger 1981; Boltho 1982; Bruno and Sachs 1985; Goldthorpe 1984; Gourevitch 1986). Again, in the vocabulary of this book, an exogenous shock modifies the actors' payoffs in such a way that cooperation becomes less likely.

Bates (1983, 14–16) offers another case of modifying the likelihood of cooperation by the payoff matrix of the game. In his reinterpretation of Evans-Pritchard's *The Nuer*, Bates insists on the existence of cross-cutting ties between tribes as cooperation-inducing mechanisms. Indeed, when a man has a wife from another group, he has a material interest in cooperation with that group because the temptation payoff is reduced when he quarrels with his wife's group: "she can make life pretty unpleasant for him." Again, the reduction of T in the payoff matrix of the game has the consequence of increasing the likelihood of cooperation (proposition 3.7).

In all these studies, the same theme reappears: some exogenous shock or different conditions account for differences in the players' payoff matrix, which matrix affects the likelihood of cooperation. All these analyses rely heavily on one important and unstated assumption: that the interaction is either iterated and information is incomplete, or the actors are able to develop correlated strategies because, as my analysis in Chapter 3 indicated, such analyses do not hold in one-shot games without correlated strategies.

3. Most Marxists, however, would deny that workers and capitalists are involved in a prisoners' dilemma game. If the best action for the working class is revolution, then the appropriate model for the interaction between capitalists and workers is the deadlock game.

In Chapter 6, I examined the converse problem: the game between elites was nested inside the principal-agent relation that each elite entertained with its followers, and this agency relationship determined the payoffs of the elites in the parliamentary game. Figure 3.3B offered a summary description of the argument in Chapter 6: the followers influence the payoffs of their respective leaders in a scheme reminiscent of the Greek letter pi.

A similar pi-like explanation can be found in the account of the failure of the Powell amendment in the U.S. House of Representatives, as presented by Denzau, Riker, and Shepsle (1985). In 1956, while a bill on federal aid to education was being discussed in the House of Representatives, a black representative from Harlem, Adam Clayton Powell, presented an amendment specifying that federal aid be given only to nonsegregated schools.[4] There is substantial evidence that the Republican leadership voted for the amendment in order to kill the bill in the final vote on the floor. Former President Truman had warned the Democratic leadership that this outcome was likely and had asked the Democrats to vote against the Powell amendment so that, at a minimum, schools (regardless of segregation) would receive federal aid (see *Congressional Record* Vol. 102, Part 9, p. 11758). His warnings were not heeded. As Truman had anticipated, the amendment was accepted, and the amended bill was then defeated. Ninety-seven votes switched from yea on the amendment to nay on the final passage, indicating a substantial amount of sophisticated voting.

According to Denzau, Riker, and Shepsle (1985), Republicans could vote strategically while liberal Democrats could not because of the relation of each group with its constituency: the liberal Democrats came from constituencies more sensitive to racial issues. The incumbents would have had a hard time explaining their sophisticated vote against the Powell amendment to their constituents back home.

Putnam (1988) adopted a similar approach in what he calls two-level games. He examines international economic summits, which present the characteristic that the agreements have to be ratified by national parliaments. Both Denzau, Riker, and Shepsle

---

4. For the details of the story and the different political groups backing the Powell amendment, the bill, and the status quo ante, see Riker (1983).

and Putnam use sequential games, as opposed to the simultaneous games of this book, in their approach: in the first round, the elites interact; in the second, the followers approve or disapprove.

Axelrod (1987) studied the interaction between domestic and international politics using what he calls the "gamma paradigm." He focused on the interaction of the American administration with the Soviet leadership, on the one hand, and American public opinion, on the other. He provided reasons he is not concerned with the relationship between Soviet leaders and the Soviet public. The outcome is that his explanatory scheme resembles the letter gamma instead of the letter pi.

Scharpf (1987) used a similar approach to investigate the political economy of four European countries in the 1970s. Lange (1984) used a conceptually similar model to investigate negotiations between labor and capital. Lange and Tsebelis (1988) used a bargaining model with incomplete information where outside actors (labor activists, government, international competition) influence the payoffs of labor and capital to account for strike activities in capitalist countries. Tsebelis (forthcoming) uses the nested games framework to investigate the impact of domestic politics on international economic sanctions. Alt and Eichengreen (1988) developed the concepts of parallel and overlapping games to investigate the issue of European gas trade.

Despite their differences, all these studies share the characteristic that one or two other parties influence the way political elites interact with each other. Again, in all these cases, the unstated assumption is that the games are iterated and information incomplete; otherwise changes in the payoffs would not necessarily change actors' behavior, as Chapter 3 demonstrated.

In Chapter 7, the prisoners' dilemma game between the partners of each coalition at the national level was nested inside a game at the constituency level, the conditions prevailing at the local level determined each player's payoffs, which, in turn, determined the likelihood of cohesion within each coalition. The summary representation of the argument is presented in Figure 3.3C.

I know of no other study that uses this particular or a similar representation. However, one can imagine that the same approach will be useful whenever two coalitions confront each other: for instance, East-West relations or labor-management interaction.[5]

Summary of findings. In Chapter 5, we saw that the frequency of disagreement between activists and the NEC did not reveal any information about the distribution of power inside the party. In single-shot games under perfect information, there is no possibility of open conflict between activists and MPs or between activists and the NEC. The conflict can arise only under incomplete information: a signalling process that indicates one actor's desire to select a different equilibrium in the reselection game. Therefore, open disagreement can be a signal from either actor and bears no relation to power. The only meaningful prediction that can be made in this context comes from the nested games framework: under the assumptions of the model, marginal constituencies are more likely to have moderate MPs than safe ones. Empirical tests of dissent in the 1974–79 House of Commons corroborated the theoretic expectations.

In Chapter 6, I used this explicit modelling of iterations between players to account for elite-initiated conflict. The elites were involved in a game of chicken, and by initiating conflict, they attempted to select the most advantageous Nash equilibrium. Neither the consociational literature nor the sophisticated voting accounts could explain this phenomenon because they avoid the use of game theoretic concepts. As a result, they cannot account for contingent behavior.

In Chapter 7, I put the nested games theory to a stricter empirical test. I found a series of empirical regularities that corroborated its predictions while falsifying those made by spatial models or other theories. The analysis of nested games also enabled me to accounted for social-psychological concepts such as *partisan identity*. In my account, the distribution of strength among the four families accounts for variations in partisan sympathies.

Substantively, nested games is a way of transplanting context into game theory. In fact, instead of assuming that people play games in a vacuum, it shows that these games are embedded in some higher order network. This network of games in my approach determines the players' payoffs.

---

5. Assuming labor and capital are composed of several organizations each.

All my results point to the conclusion that political context matters in ways that are predictable because they influence different actors' payoffs in nested games, and these payoffs influence the choice of strategies.

III. Political Institutions

Rational-choice literature usually regards institutions as constraints on the actions of rational actors. Chapter 4 investigated the converse problem: the problem of institutional design. This discussion represents an exploration of uncharted territory. Arguably the most important contribution of this chapter was its classification of institutions into efficient and redistributive. In addition, Chapter 4 provided a classification of redistributive institutions into consolidating and new deal institutions. This particular distinction is important because previous accounts merely extrapolated from one of these categories. For economists, institutions are efficient; for Marxists, they are consolidating; for liberals, they are new deal. I argue that it is more fruitful to recognize different kinds of institutions, speculating that lack of information about future events is likely to produce more efficient than redistributive institutions, whereas perfect information about future events will produce redistributive institutions either of the consolidating or of the new deal variety.

The empirical chapters of the book considered institutions as exogenously given and studied human behavior inside the existing institutional framework. However, in each chapter, I demonstrated why some or all of the actors were not satisfied with the outcomes produced by existing institutions. This observation led to the study of institutional change as deliberate political activity.

In Chapter 5, the game theoretic analysis of Labour party institutions demonstrated that contrary to existing beliefs, the mandatory reselection clause did not fundamentally alter the balance of forces between Right and Left or party constituencies and leadership. Rather, it reflected the change in the balance of forces that resulted from the shift of the trade unions to the left during the 1970s. This modest institutional reform was the best the constituency activists could do because this was the only reform that the trade unions would support. The new Labour party constitution was an example of a redistributive institution of the new deal variety because it realigned part of the previous winning coalition (the trade unions) with the previous losers (the constituency activists) in order to redistribute power inside the party.

In Chapter 6, I showed that institutions are more likely than vote trading to produce consociational outcomes. The very existence of consociational institutions indicated that vote trading across asymmetric issues was not a sufficient solution for conflict management in consociational democracies. We can say that the role of institutions is exactly the opposite of that of facilitating vote trading: institutions assign exclusive jurisdictions on issues of asymmetric importance. In so doing, Belgian institutions push outcomes toward the Pareto frontier and are efficient.

Chapter 7 demonstrated how institutions can be used as resources in the political battle and how redistributive institutions of the consolidating variety can be created. Each coalition in power selected an electoral system that would improve its position at the expense of the loser. Moreover, the element of uncertainty was strategically manipulated through the use of article 38 of the constitution so that the outcome would be efficient for the members of the majority coalition.

The politics of institutional change was described with very few details in Chapter 6, in which the adopted institutions pushed outcomes toward the Pareto frontier. They were, however, treated more extensively in Chapters 5 and 7, in which institutional change was part of the political conflict itself. In both Chapters 5 and 7, I had to explain the specific institutional changes as optimal outcomes, that is, show that they were better than previous existing outcomes and that no other feasible option would make a winning coalition better off.

My classification of institutions and my conjectures about the informational preconditions of efficient and redistributive institutions are, however, far from being theoretical statements. Theoretical investigation of the issues and systematic tests of the theory are required.

IV. A Final Note

The success of the nested games framework in each of the topics treated in this book varied according to the quality of the data as well as with the existence of well-developed theories. This success should not distract readers, however, from the most important achievement: it is essentially the same theory that is tested in each different setting.

The core of the theory consists of some very simple ideas: seemingly suboptimal choices indicate the presence of nested games (either games in multiple arenas or institutional design); in games in multiple arenas, events or strategies in one arena influence the way the game is played in another arena; institutional design refers to the choice of rules as opposed to the choice of strategies inside existing rules. The combination of these ideas was sufficient to explain a variety of phenomena, and the generality of the treatment indicates that the applications can be multiplied.

The phenomena covered here do not by any means exhaust the list of applications of nested games or of institutional design. In the introduction and the conclusions of Chapters 5, 6, and 7, I indicated possible applications of nested games. They include coalition theory, class conflict, factional politics, international/domestic politics issues, reputation, ideology, and the study of institutions.

This book has demonstrated ways of using the idea of nested games in only three areas of comparative Western European politics: institution building, party politics, and coalitions. It is not meant to be the last word on any of these issues. It strives to be one of the first words in a growing approach to comparative politics.

Bibliography

Abel, T. 1948. "The Operation Called *Verstehen*." *American Journal of Sociology* 54:211–18.

Abrams, R. 1980. *Foundations of Political Analysis*. New York: Columbia University Press.

Abreu, D. 1986. "Extremal Equilibria of Oligopolistic Supergames." *Journal of Economic Theory* 39:191–225.

Alchian, A. A. 1984. "Specificity, Specialization and Coalitions." *Journal of Institutional and Theoretical Economics* 140:34–49.

Aldrich, J. H. 1983a. "A Downsian Spatial Model with Party Activism." *American Political Science Review* 77:974–90.

———. 1983b. "A Spatial Model with Party Activists: Implications for Electoral Dynamics." *Public Choice* 41:63–99.

Allum, P. A. 1973. *Italy—Republic Without Government?* New York: W. W. Norton.

Almond, G. 1956. "Comparative Political Systems." *Journal of Politics* 18:391–409.

Almond, G., and Verba, S. 1963. *The Civic Culture: Political Attitudes and Democracy in Five Nations*. Princeton, N.J.: Princeton University Press.

Alt, J. E., and Eichengreen, B. 1988. "Parallel and Overlapping Games: Theory and an Application to the European Gas Trade." Mimeo.

Apter, D. A. 1965. *The Politics of Modernization*. Chicago: University of Chicago Press.

Arrow, K. J. 1951. *Social Choice and Individual Values*. New Haven, Conn.: Yale University Press.

———. 1965. *Aspects of the Theory of Risk-Bearing*. Helsinki: Yrjo Jahnsson Saatio.

Aumann, R. J. 1974. "Subjectivity and Correlation in Randomized Experiments." *Journal of Mathematical Economics* 1:67–96.

Austen-Smith, D., and Banks, J. 1988. "Elections, Coalitions, and Legislative Outcomes." *American Political Science Review* 82:405–22.

Avril, P., and Gicquel, J. 1985. "Chronique Constitutionnelle Française." *Pouvoirs* 35:175–99.

———. 1986. "Chronique Constitutionnelle Française." *Pouvoirs* 39:159–79.

Axelrod, R. 1970. *Conflict of Interest, A Theory of Divergent Goals with Applications to Politics*. Chicago: Markham Press.

———. 1981. "The Emergence of Cooperation Among Egoists." *American Political Science Review* 75:306–18.

———. 1984. *The Evolution of Cooperation*. New York: Basic Books.

———. 1987. "The Gamma Paradigm for Studying the Domestic Influence on Foreign Policy." Mimeo.

Axelrod, R., and Hamilton, W. D. 1981. "The Evolution of Cooperation." *Science* 211:1390–96.

Bacharach, M. 1987. "A Theory of Rational Decision in Games." *Erkenntnis* 27:17–55.

Banting, G., and Simeon, R. 1985. *Redesigning the State: The Politics of Constitutional Change*. Toronto: University of Toronto Press.

Barry, B. 1975a. "The Consociational Model and Its Dangers." *European Journal of Political Research* 3:393–412.

———. 1975b. "Review Article: Political Accommodation and Consociational Democracy." *British Journal of Political Science* 5:477–505.

———. 1978. *Sociologists, Economists, and Democracy*. Chicago: University of Chicago Press.

Barry, B., and Hardin, R., eds. 1982. *Rational Man, Irrational Society?* Beverly Hills, Calif.: Sage Publications.

Bartolini, S. 1984. "Institutional Constraints and Party Competition in the French Party System." *West European Politics* 7:103–27.

Bates, R. H. 1974. "Ethnic Competition and Modernization in Contemporary Africa." *Comparative Political Studies* 7:457–89.

———. 1982. "Modernization, Ethnic Competition, and the Rationality of Politics in Africa." In D. Rotchild and V. Olorunsula (eds.), *Ethnic Conflict in Africa*. Boulder, Colo.: Westview Press.

———. 1983. *Essays on the Political Economy of Rural Africa*. Cambridge: Cambridge University Press.

———. 1985. "The Analysis of Institutions." Paper presented to the Carnegie Conference on the Study of Institutions, Pittsburgh.

Beer, S. H. 1969. *Modern British Politics*. London: Faber and Faber.

Bendix, R. 1964. *Nation-Building and Citizenship*. New York: Wiley.

Bendor, J., and Mookherjee, D. 1987. "Institutional Structure and the

Logic of Ongoing Collective Action." *American Political Science Review* 81:129–54.

Bentley, A. F. 1908. *The Process of Government.* Chicago: University of Chicago Press.

Berger, S., ed. 1981. *Organizing Interests in Western Europe.* Cambridge: Cambridge University Press.

Bernheim, B. D. 1984. "Rationalizable Strategic Behaviour." *Econometrica* 52:1007–28.

Bhaduri, A. 1976. "A Study of Agricultural Backwardness Under Semi-Feudalism." *Economic Journal* 83:120–37.

Billiet, J. 1984. "On Belgian Pillarization: Changing Perspectives." *Acta Politica* 19:117–28.

Binmore, K. 1987. "Modeling Rational Players: Part I." *Economics and Philosophy* 3:179–214.

Bochel, J., and Denver, D. 1983. "Candidate Selection in the Labour Party: What the Selectors Seek." *British Journal of Political Science* 13:45–69.

Boltho, A. 1982. *The European Economy: Growth and Crisis.* Oxford: Oxford University Press.

Bon, F. 1985. "Rétrosimulations Proportionalistes." *Pouvoirs* 32:135–49.

Bonanno, G. 1988. "The Logic of Rational Play in Sequential Games." Mimeo, University of California, Davis.

Borella, F. 1984. *Les Partis Politiques en Europe.* Paris: Seuil.

Boudon, R. 1977. *Effet Perverse et Ordre Social.* Paris: Presses Universitaires de France.

———. 1979. *La Logique du Social.* Paris: Librairie Hachette.

———. 1984. *La Place du Désordre.* Paris: Presses Universitaires de France.

———. 1986. *L'Idéologie.* Paris: Fayard.

Bourdieu, P. 1979. *La Distinction.* Paris: Minuit.

Bradley, I. 1981. *Breaking the Mould?* Oxford: Martin Robinson.

Brady, D. W. 1973. *Congressional Voting in a Partisan Era: A Study of the McKinley Houses.* Lawrence: University of Kansas Press.

Brams, S. J. 1975. "Newcomb's Problem and Prisoners' Dilemma." *Journal of Conflict Resolution* 19:596–612.

Bruno, M., and Sachs, J. 1985. *Economics and Worldwide Stagflation.* Cambridge, Mass.: Harvard University Press.

Butler, D. E. 1960. "The Paradox of Party Difference." *American Behavioral Scientist* 4:3–5.

Butler, D. E., and Kavanagh, D. 1980. *The British General Election of 1979.* London: Macmillan.

————. 1984. *The British General Election of 1983.* London: Macmillan.

Butler, D. E., and Pinto-Duschinsky, M. 1971. *The British General Election of 1970.* London: Macmillan.

Cain, B., Ferejohn, J., and Fiorina, M. 1987. *The Personal Vote: Constituency Service and Electoral Independence.* Cambridge, Mass.: Harvard University Press.

Calvert, R. L. 1985. "Uncertainty, Asymmetry, and Reciprocity in Repeated Two-Player Games." Paper presented at the American Political Science Association Conference, Chicago.

Calvert, R. L., and McKuen, M. 1985. "Bayesian Learning and the Dynamics of Public Opinion." Paper presented at the Midwest Political Science Association Conference, Chicago.

Cameron, D. R. 1978. "The Expansion of the Public Economy: A Comparative Analysis." *American Political Science Review* 72: 1243–61.

Capdevielle, J., Dupoivier, E., Grunberg, G., Schweisguth, E., and Ysmal, C. 1981. *France de Gauche Vote A Droite.* Paris: Presses de la Fondation Nationale des Sciences Politiques.

Carroll, J. W. 1985. "Indefinite Terminating Points and the Iterated Prisoners' Dilemma." Mimeo, University of Arizona.

Cayrol, R., Flavigny, P. O., and Fournier, I. 1981. "Que Donnerait la Proportionnelle?" *Le Monde*, August 19, p. 2.

Chagnollaud, D. 1985. "Les Presidents de la Ve Republique et le Mode d'Election des Deputés a l'Assemblée Nationale." *Pouvoirs* 32: 95–118.

Chapsal, J., and Lancelot, A. 1969. *La Vie Politique en France Depuis 1940.* Paris: Presses Universitaires de France.

Chew, S. H. 1983. "A Generalization of the Quasilinear Mean with Applications to the Measurement of Income Inequality and Decision Theory Resolving the Allais Paradox." *Econometrica* 51: 1065–92.

Converse, P. E. 1969. "Of Time and Partisan Stability." *Comparative Political Studies* 2: 139–70.

Converse, P. E., and Pierce, R. 1986. *Political Representation in France.* Cambridge, Mass.: Harvard University Press.

Coser, L. 1971. "Social Conflict and the Theory of Social Change." In C. G. Smith (ed.), *Conflict Resolution: Contributions of the Behavioral Sciences.* Notre Dame, Ind.: University of Notre Dame Press.

Covell, M. 1981. "Ethnic Conflict and Elite Bargaining: The Case of Belgium." *West European Politics* 4: 197–218.

————. 1982. "Agreeing to Disagree: Elite Bargaining and the Revision of the Belgian Constitution." *Canadian Journal of Political Science* 15: 451–69.

Craig, F. W. S. 1984. *British Parliamentary Election Results 1974–1983.* West Sussex, England: Parliamentary Research Services.

Crewe, I. 1981. "Electoral Participation." In D. Butler, H. R. Penniman, and A. Ranney (eds.), *Democracy at the Polls.* Washington, D.C.: American Enterprise Institute.

Criddle, B. 1984. "Candidates." In D. Butler and D. Kavanagh (eds.), *The British General Election of 1983.* London: Macmillan.

CRISP (Centre de Recherche et d'Information Socio-Politique). 1983. *Le Systéme de la Décision Politique en Belgique.* Bruxelles: CRISP.

Daalder, H. 1966. "The Netherlands: Opposition in a Segmented Society." In R. A. Dahl (ed.), *Political Oppositions in Western Democracies.* New Haven, Conn.: Yale University Press.

Dahl, R. A. 1956. *A Preface to Democratic Theory.* Chicago: University of Chicago Press.

Dallmayr, F. R., and McCarthy, T. A. 1977. *Understanding and Social Inquiry.* Notre Dame, Ind.: University of Notre Dame Press.

Davidson, D., McKinsey, J. C. C., and Suppes, P. 1954. "Outline of a Formal Theory of Value, I." *Philosophy of Science* 22:140–60.

Denis, N. 1978. "Les Elections Législatives de Mars 1978 en Métropole." *Revue Française de Science Politique* 28:977–1005.

Denzau, A., Riker, W., and Shepsle, K. 1985. "Farquharson and Fenno: Sophisticated Voting and Home Style." *American Political Science Review* 79:1117–34.

De Ridder, M., and Fraga, L.-R. 1986. "The Brussels Issue in Belgian Politics." *West European Politics* 9:376–92.

De Ridder, M., Peterson, R. L., and Wirth, R. 1978. "Images of Belgian Politics: The Effects of Cleavages on the Political System." *Legislative Studies Quarterly* 3:83–108.

Dickson, A. D. R. 1975. "MP's Readoption Conflicts: Their Causes and Consequences." *Political Studies* 23:62–70.

Dierickx, G. 1978. "Ideological Oppositions and Consociational Attitudes in the Belgian Parliament." *Legislative Studies Quarterly* 3:133–60.

DiPalma, G. 1976. "Institutional Rules and Legislative Outcomes in the Italian Parliament." *Legislative Studies Quarterly* 1:147–80.

Dodd, L. 1976. *Coalitions in Parliamentary Government.* Princeton, N.J.: Princeton University Press.

Downs, A. 1957. *An Economic Theory of Democracy.* New York: Harper and Row.

Duverger, M. 1954. *Political Parties.* New York: Wiley. (French edition 1952.)

———. 1968. *Institutions Politiques et Droit Constitutionnel,* 10th ed.

Paris: Presses Universitaires de France.

———. 1978. *Echec au Roi*. Paris: A. Michel.

———. 1982. *La République des Citoyens*. Paris: Ramsay.

Easton, D. 1953. *The Political System*. New York: Knopf.

———. 1957. "An Approach to the Analysis of Political Systems." *World Politics* 11:383–400.

Edwards, W. 1954. "The Theory of Decision Making." *Psychological Bulletin* 51:380–417.

Elster, J. 1978. *Logic and Society*. New York: Wiley.

———. 1983. *Explaining Technical Change*. Cambridge: Cambridge University Press.

———. 1985. *Making Sense of Marx*. Cambridge: Cambridge University Press.

Epstein, L. D. 1960. "British M.P.s and Their Local Parties: The Suez Cases." *American Political Science Review* 54:374–90.

———. 1967. *Political Parties in Western Democracies*. New York: Praeger.

Farquharson, R. 1969. *Theory of Voting*. New Haven, Conn.: Yale University Press.

Fenno, R. F. 1978. *Home Style*. Boston: Little, Brown.

Ferejohn, J. A. 1974. *Pork Barrel Politics*. Stanford, Calif.: Stanford University Press.

Ferejohn, J. A., and Fiorina, M. P. 1974. "The Paradox of Non-Voting: A Decision-Theoretic Analysis." *American Political Science Review* 68:525–36.

Finer, S. E. 1981. *The Changing British Party System, 1945–1979*. Washington, D.C.: American Enterprise Institute.

Fiorina, M. P. 1974. *Representatives, Roll Calls, and Constituencies*. Lexington, Mass.: Lexington Books.

———. 1981. *Retrospective Voting in American Elections*. New Haven, Conn.: Yale University Press.

Fishburn, P. C. 1983. "Transitive Measurable Utility." *Journal of Economic Theory* 31:293–317.

Flood, M. M. 1952. "Some Experimental Games." Research memorandum RM-789. Santa Monica, Calif.: RAND Corporation.

Friedman, J. 1971. "A Noncooperative Equilibrium for Supergames." *Review of Economic Studies* 38:1–12.

———. 1977. *Oligopoly and the Theory of Games*. Amsterdam: North-Holland.

Friedman, M. 1953. "The Methodology of Positive Economics." In M. Friedman (ed.), *Essays in Positive Economics*. Chicago: University of Chicago Press.

———. 1962. *Capitalism and Freedom*. Chicago: University of Chicago Press.

Fudenberg, D., and Maskin, E. 1986. "The Folk Theorem in Repeated Games with Discounting or with Incomplete Information." *Econometrica* 54:533–54.

Gauthier, D. P. 1986. *Moral by Agreement*. Oxford: Oxford University Press.

Gibbard, A. 1973. "Manipulation of Voting Schemes: A General Result." *Econometrica* 41:587–601.

Goldman, S. M. 1969. "Consumption Behavior and Time Preference." *Journal of Economic Theory* 1:39–47.

Goldthorpe, J., ed. 1984. *Order and Conflict in Contemporary Capitalism*. Oxford: Clarendon Press.

Gourevitch, P. 1986. *Politics in Hard Times*. Ithaca, N.Y.: Cornell University Press.

Grofman, B. 1982. "A Dynamic Model of Protocoalition Formation in Ideological N-Space." *Behavioral Science* 27:77–90.

Gurr, T. R. 1971. *Why Men Rebel*. Princeton, N.J.: Princeton University Press.

Halpern, S. M. 1986. "The Disorderly Universe of Consociational Democracy." *West European Politics* 9:181–97.

Haltiwanger, J., and Waldman, M. 1985. "Rational Expectations and the Limits of Rationality: An Analysis of Heterogeneity." *American Economic Review* 75:326–40.

Hamilton, A., Madison, J., and Jay, J. 1961. *The Federalist Papers*. New York: New American Library.

Hammond, T. H., and Miller, G. J. 1987. "The Core of the Constitution." *American Political Science Review* 81:1155–74.

Hanushek, E. R., and Jackson, J. E. 1977. *Statistical Methods for Social Scientists*. New York: Academic Press.

Hardin, R. 1971. "Collective Action as an Agreeable N-Prisoners' Dilemma." *Behavioral Science* 16:472–79.

———. 1982. *Collective Action*. Baltimore: Johns Hopkins University Press.

Harsanyi, J. C. 1967–68. "Games with Incomplete Information Played by 'Bayesian Players,'" 3 parts. *Management Science* 14:159–82, 320–34, 486–502.

———. 1975. "Can the Maximin Principle Serve as a Basis for Morality? A Critique of John Rawls' Theory." *American Political Science Review* 69:594–606.

———. 1977. "Rule Utilitarianism and Decision Theory." *Erkenntnis* 11:25–53.

Hayek, F. A. 1955. *The Counterrevolution of Science*. New York: Free Press.

———. 1973, 1976, 1979. *Law, Legislation and Liberty*, 3 vols. Chicago: University of Chicago Press.

Head, J. G. 1972. "Public Goods: The Polar Case." In R. M. Bird and J. G. Head (eds.), *Modern Fiscal Issues*. Toronto: University of Toronto Press.

Hedlund, R. D. 1984. "Organizational Attributes of Legislatures: Structure, Rules, Norms, Resources." *Legislative Studies Quarterly* 9:51–121.

Heisler, M. O. 1973. "Institutionalizing Societal Cleavages in a Cooptive Polity: The Growing Importance of the Output Side in Belgium." In M. O. Heisler (ed.), *Politics in Europe: Structures and Processes in Some Postindustrial Democracies*. New York: McKay.

Hempel, C. G. 1964. *Aspects of Scientific Explanation*. New York: Free Press.

Holloway, J., and Picciotto, S. 1978. *State and Capital: A Marxist Debate*. London: Edward Arnold.

Holt, R. 1967. "A Proposed Structural-Functional Framework." In J. Charlesworth (ed.), *Contemporary Political Analysis*. New York: Free Press.

Howard, N. 1971. *Paradoxes of Rationality*. Cambridge, Mass.: MIT Press.

Howe, R. E., and Roemer, J. E. 1981. "Rawlsian Justice as the Core of a Game." *American Economic Review* 71:880–95.

Huntington, S. P. 1968. *Political Order in Changing Societies*. New Haven, Conn.: Yale University Press.

Huyse, L. 1984. "Pillarization Reconsidered." *Acta Politica* 19:145–58.

Jaffré, J. 1980. "The French Electorate in March 1978." In H. R. Penniman (ed.), *The French National Assembly Elections of 1978*. Washington, D.C.: AEI Studies.

Janosik, E. G. 1968. *Constituency Labour Parties in Britain*. New York: Praeger.

Jeffery, R. C. 1974. "Preference Among Preferences." *Journal of Philosophy* 71:377–91.

Jensen, M., and Meckling, W. 1976. "Theory of the Firm: Managerial Behavior, Agency Costs, and Ownership Structure." *Journal of Financial Economics* 3:305–60.

Jervis, R. 1978. "Cooperation Under the Security Dilemma." *World Politics* 30:167–214.

Johnson, D. B., and Gibson, J. R. 1974. "The Divisive Primary Revisited:

Party Activists in Iowa." *American Political Science Review* 68:67–77.

Kadane, J. B., and Larkey, P. D. 1982. "Subjective Probability and the Theory of Games." *Management Science* 28:113–20.

Kahneman, D., and Tversky, A. 1979. "Prospect Theory: An Analysis of Decision Under Risk." *Econometrica* 47:263–91.

Kahneman, D., Slovic, P., and Tversky, A. 1984. *Judgment Under Uncertainty*. Cambridge: Cambridge University Press.

Kalai, E., and Stanford, W. 1988. "Finite Rationality and Interpersonal Complexity in Repeated Games." *Econometrica* 56:397–410.

Karmarkar, U. S. 1978. "Subjectively Weighted Utility: A Descriptive Extension of the Expected Utility Model." *Organizational Behavior and Human Performance* 21:61–72.

Katzenstein, P. J. 1985. *Small States in World Markets*. Ithaca, N.Y.: Cornell University Press.

Kautsky, J. 1971. *The Political Consequences of Modernization*. New York: R. E. Krieger.

Keech, W. R. 1972. "Linguistic Diversity and Political Conflict: Some Observations Based on Four Swiss Cantons." *Comparative Politics* 4:387–404.

Keohane, R. O. 1981. "The Theory of Hegemonic Stability and Changes in International Economic Regimes, 1967–1977." In O. R. Holsti, A. George, and R. M. Siverson (eds.), *Change in the International System*. Boulder, Colo.: Westview Press.

Key, V. O., Jr. 1958. *Politics, Parties and Pressure Groups*, 4th ed. New York: Crowell.

———. 1968. *The Responsible Electorate*. New York: Vintage.

Keynes, J. M. 1891. *The Scope and Method of Political Economy*. London: Macmillan.

Kiewiet, R. D. 1983. *Macroeconomics and Micropolitics*. Chicago: University of Chicago Press.

Klein, B., Crawford, R., and Alchian, A. 1978. "Vertical Integration, Appropriable Rents, and the Competitive Contracting Process." *Journal of Law and Economics* 21:297–326.

Knapp, A. 1987. "Proportional but Bipolar: France's Electoral System in 1986." *West European Politics* 10:89–114.

Koelble, T. A. 1987. "Trade Unionists, Party Activists and Politicians: The Struggle for Power over Party Rules in the British Labour Party and the West German Social Democratic Party." *Comparative Politics* 19:253–66.

Kogan, D., and Kogan, M. 1982. *The Battle for the Labour Party*. New

York: St. Martin's Press.

Kohlberg, E., and Mertens, J.-F. 1986. "On the Strategic Stability of Equilibria." *Econometrica* 54:1003–37.

Kornberg, A. 1967. *Canadian Legislative Behavior: The Study of the Twenty-Fifth Parliament*. New York: Holt.

Kreps, D. M., and Wilson, R. 1982a. "Reputation and Imperfect Information." *Journal of Economic Theory* 27:253–79.

———. 1982b. "Sequential Equilibria." *Econometrica* 50:863–94.

Kreps, D. M., Milgrom, P., Roberts, J., and Wilson, R. 1982. "Rational Cooperation in the Finitely Repeated Prisoners' Dilemma." *Journal of Economic Theory* 27:245–52.

Kuhn, T. S. 1962. *The Structure of Scientific Revolutions*. Chicago: University of Chicago Press.

Labour Party. 1970. *Report of the Annual Conference.*

———. 1979. *Report of the Annual Conference.*

———. 1984–85. *Rule Book.*

Lakatos, I. 1970. "The Methodology of Scientific Research Programs." In I. Lakatos and A. Musgrave (eds.), *Criticism and the Growth of Knowledge*. Cambridge: Cambridge University Press.

Lange, P. 1984. "Unions, Workers and Wage Regulation: The Rational Bases of Consent." In J. Goldthorpe (ed.), *Order and Conflict in Contemporary Capitalism*. Oxford: Clarendon Press.

———. 1987. "Institutionalization of Concertation." Duke University Program for International Political Economy Paper #26.

Lange, P., and Tsebelis, G. 1988. "Wage Negotiations and Strikes: An Equilibrium Analysis." Paper presented at the American Political Science Association meeting, Washington, D.C.

Lavau, G., and Mossuz-Lavau, J. 1980. "The Union of the Left's Defeat: Suicide or Congenital Weakness?" In H. R. Penniman (ed.), *The French National Assembly Elections of 1978*. Washington, D.C.: AEI Studies.

Laver, M. 1977. "Intergovernmental Policy on Multinational Corporations, a Simple Model of Tax Bargaining." *European Journal of Political Research* 5:363–80.

LeGall, G. 1985. "Loi Electorale: Dégâts et Réflections Autour d'une Réforme." *Revue Politique et Parlementaire* 918:27–34.

Lehmbruch, G. 1974. "Consociational Democracy in the International System." *European Journal of Political Research* 3:377–91.

Le Monde (Dossiers et Documents). *Les Elections Législatives de Mars 1978*. Paris.

Levi, I. 1980. *The Enterprise of Knowledge: An Essay on Knowledge, Credal Probability, and Chance*. Cambridge, Mass.: MIT Press.

Lijphart, A. 1968. *The Politics of Accommodation: Pluralism and Democracy in the Netherlands.* Berkeley: University of California Press.

———. 1969. "Consociational Democracy." *World Politics* 21:207–25.

———. 1977. *Democracy in Plural Societies.* New Haven, Conn.: Yale University Press.

Lindblom, C. E. 1977. *Politics and Markets.* New York: Basic Books.

Long, N. 1961. *The Polity.* Chicago: Rand McNally.

Lorwin, V. R. 1966. "Belgium: Religion, Class, and Language in National Politics." In R. A. Dahl (ed.), *Political Oppositions in Western Democracies.* New Haven, Conn.: Yale University Press.

———. 1971. "Segmented Pluralism: Ideological Cleavages and Political Cohesion in the Smaller European Democracies." *Comparative Politics* 3: 141–75.

Lucas, G. 1982. *Studies in Business-Cycle Theory.* Cambridge, Mass.: MIT Press.

Luce, R. D., and Raiffa, H. 1957. *Games and Decisions.* New York: Wiley.

Luebbert, G. 1983. "Coalition Theory and Government Formation in Multiparty Democracies." *Comparative Politics* 15:235–49.

Mabille, X. 1986. *Histoire Politique de la Belgique.* Bruxelles: Editions du Centre de Recherche et d'Information Socio-Politiques.

Machina, M. J. 1982. "'Expected Utility' Analysis Without the Independence Axiom." *Econometrica* 50:277–323.

Macrae, D., Jr. 1958. *Dimensions of Congressional Voting.* Berkeley: University of California Press.

March, J. 1978. "Bounded Rationality, Ambiguity, and the Engineering of Choice." *Bell Journal of Economics* 9:587–608.

Marx, K. 1963. *Early Writings,* ed. T. Bottomore. London: McGraw-Hill.

———. 1967. *Capital.* New York: International Publishers.

May, J. D. 1969. "Democracy, Party 'Evolution,' Duverger." *Comparative Political Studies* 2:217–48.

———. 1973. "Opinion Structure in Political Parties: The Special Law of Curvilinear Disparity." *Political Studies* 11:135–51.

Mayhew, D. 1966. *Party Loyalty Among Congressmen.* Cambridge, Mass.: Harvard University Press.

Maynard Smith, J. 1982. *Evolution and the Theory of Games.* Cambridge: Cambridge University Press.

McCormick, P. 1981. "Prentice and the Newham North-East Constituency: The Making of Historical Myths." *Political Studies* 29:73–90.

McKelvey, R. D. 1976. "Intransitivities in Multidimensional Voting Models and Some Implications for Agenda Control." *Journal of Economic Theory* 12:472–82.

———. 1979. "General Conditions for Global Intransitivities in Formal Voting Models." *Econometrica* 47:1085–1112.

McKenzie, R. T. 1964. *British Political Parties*, 2nd ed. New York: Praeger.

———. 1982. "Power in the Labour Party: The Issue of Intra-Party Democracy." In D. Kavanagh (ed.), *The Politics of the Labour Party*. London: Allen and Unwin.

McLean, I. 1978. "Labour Since 1945." In C. Cook and J. Ramsden (eds.), *Trends in British Politics Since 1945*. London: Macmillan.

McRae, K. 1974. *Consociational Democracy: Political Accommodation in Segmented Societies*. Toronto: McClelland and Stewart.

Meyer, F. 1978. "Et Si la France Avait Voté à la Proportionnelle?" *Le Matin de Paris, Le Dossier des Législatives*.

Michels, R. 1949. *Political Parties*. Glencoe, Ill.: Free Press.

Milgrom, P., and Roberts, J. 1982a. "Limit Pricing and Entry Under Incomplete Information: An Equilibrium Analysis." *Econometrica* 50:443–59.

———. 1982b. "Predation, Reputation, and Entry Deterrence." *Journal of Economic Theory* 27:280–312.

Miliband, R. 1961. *Parliamentary Socialism*. London: Merlin Press.

Miller, G. J., and Moe, T. M. 1983. "Bureaucrats, Politicians, and the Size of Government." *American Political Science Review* 77:297–322.

Minkin, L. 1978. *The Labour Party Conference*. London: Allen Lane.

Mosca, G. 1939. *The Ruling Class*, ed. A. Livingston. New York: McGraw-Hill.

Moulin, H. 1982. *Game Theory for the Social Sciences*. New York: New York University Press.

Musgrave, A. 1981. "'Unreal Assumptions' in Economic Theory: The F-Twist Untwisted." *Kyklos* 34:377–87.

Muth, J. 1961. "Rational Expectations and the Theory of Price Movements." *Econometrica* 29:315–35.

Myerson, R. B. 1978. "Refinements of the Nash Equilibrium Concept." *International Journal of Game Theory* 7:73–80.

———. 1987. "An Introduction to Game Theory." In S. Reiter (ed.), *Studies in Mathematical Economics*. Washington, D.C.: Mathematical Association of America.

Nagel, E. 1963. "Assumptions in Economic Theory." *American Economic Review Papers and Proceedings* 53:211–19.

Nash, J. F., Jr. 1951. "Noncooperative Games." *Annals of Mathematics* 54:289–95.

Nelson, R., and Winter, S. 1982. *An Evolutionary Theory of Economic Change.* Cambridge, Mass.: Harvard University Press.

Neumann, M. 1985. "Long Swings in Economic Development, Social Time Preference and Institutional Change." *Journal of Institutional and Theoretical Economics* 141:21–35.

Norton, P. 1975. *Dissension in the House of Commons 1945–74.* London: Macmillan.

———. 1980. *Dissension in the House of Commons 1974–1979.* Oxford: Oxford University Press.

———. 1984. "Britain: Still a Two-Party System?" *West European Politics* 7:27–45.

Nozick, R. 1974. *Anarchy, State, and Utopia.* New York: Basic Books.

Offe, C., and Wiesenthal, H. 1980. "Two Logics of Collective Action: Theoretical Notes on Social Class and Organizational Form." *Political Power and Social Theory* 1:67–115.

Olson, M., Jr. 1965. *The Logic of Collective Action.* Cambridge, Mass.: Harvard University Press.

Ordeshook, P. C. 1986. *Game Theory and Political Theory.* Cambridge: Cambridge University Press.

Ostrogorski, M. 1892. *Democracy and the Organization of Political Parties.* London: Macmillan.

Oye, K. A., ed. 1986. *Cooperation Under Anarchy.* Princeton, N.J.: Princeton University Press.

Parodi, J. L. 1983. "La Cinquième République à d'Epreuve de la Proportionelle." *Revue Française de Science Politique* 33:987–1008.

———. 1985. "La Représentation Proportionelle et le Systeme Institutionnel." *Pouvoirs* 32:43–51.

Parsons, T. 1951. *The Social System.* New York: Free Press.

Paterson, P. 1967. *The Selectorate.* London: MacGibbon and Kee.

Pearce, D. G. 1984. "Rationalizable Strategic Behaviour and the Problem of Perfection." *Econometrica* 52:1029–50.

Poggi, G. 1978. *The Development of the Modern State.* Stanford, Calif.: Stanford University Press.

Popkin, S. L. 1979. *The Rational Peasant.* Berkeley: University of California Press.

Popper, K. R. 1962. *Conjectures and Refutations: The Growth of Scientific Knowledge.* New York: Basic Books.

Powell, G. B. 1982. *Contemporary Democracies.* Cambridge, Mass.: Harvard University Press.

Pratt, J. W. 1964. "Risk Aversion in the Small and in the Large." *Econometrica* 32:122–36.

Prewitt, K., and Stone, A. 1973. *The Ruling Elites: Elite Theory, Power and American Democracy*. New York: Harper and Row.

Przeworski, A. 1980. "Material Bases of Consent: Politics and Economics in a Hegemonic System." *Political Power and Social Theory* 1:23–68.

———. 1985. "The Challenge of Methodological Individualism to Marxist Analysis." In P. Birnbaum and J. Leca (eds.), *Sur l' Individualisme*. Paris: Presse de la FNSP.

———. 1986. "Some Problems in the Study of the Transition to Democracy." In G. O'Donnell et al. (eds.), *Transitions from Authoritarian Rule*. Baltimore: Johns Hopkins University Press.

Przeworski, A., and Teune, H. 1970. *The Logic of Comparative Social Inquiry*. New York: Wiley.

Przeworski, A., and Wallerstein, M. 1982. "The Structure of Class Conflict in Democratic Capitalistic Societies." *American Political Science Review* 76:215–38.

———. 1988. "Structural Dependence of State on Capital." *American Political Science Review* 82:11–30.

Putnam, R. D. 1988. "Diplomacy and Domestic Politics: The Logic of Two-Level Games." *International Organization* 42:427–60.

Pye, L., and Verba, S., eds. 1965. *Political Culture and Political Development*. Princeton, N.J.: Princeton University Press.

Ranney, A. 1965. *Pathways to Parliament: Candidate Selection in Britain*. Madison: University of Wisconsin Press.

———. 1968. "Candidate Selection and Party Cohesion in Britain and the United States." In W. J. Crotty (ed.), *Approaches to the Study of Party Organization*. Boston: Allyn and Bacon.

Rapoport, A. 1960. *Fights, Games and Debates*. Ann Arbor: University of Michigan Press.

———, ed. 1974. *Game Theory as a Theory of Conflict Resolution*. Dordrecht, Holland: D. Reidel.

———. 1975. "Comment on Brams's Discussion of Newcomb's Paradox." *Journal of Conflict Resolution* 19:613–19.

Rawls, J. 1971. *A Theory of Justice*. Cambridge, Mass.: Harvard University Press.

Rice, P. M. 1976. "Local Strategies and Equilibria with an Application to the Committee Decision Process." *International Journal of Game Theory* 8:1–12.

Riker, W. H. 1962. *The Theory of Political Coalitions*. New Haven, Conn.: Yale University Press.

————. 1980. "Implications from the Disequilibrium of Majority Rules for the Study of Institutions." *American Political Science Review* 74:432–47.

————. 1982. "The Two-party System and Duverger's Law: An Essay on the History of Political Science." *American Political Science Review* 76:753–66.

————. 1983. *Liberalism Against Populism*. San Francisco: W. H. Freeman.

————. 1986. *The Art of Political Manipulation*. New Haven: Yale University Press.

Riker, W. H., and Ordeshook, P. C. 1968. "A Theory of the Calculus of Voting." *American Political Science Review* 62:25–42.

————. 1973. *An Introduction to Positive Political Theory*. Englewood Cliffs, N.J.: Prentice-Hall.

Rochon, T. R., and Pierce, R. 1985. "Coalitions as Rivalries: French Socialists and Communists, 1967–1978." *Comparative Politics* 17:437–51.

Rose, R. 1962. "The Political Ideas of English Party Activists." *American Political Science Review* 56:360–71.

Rosenthal, H., and Sen, S. 1973. "Electoral Participation in the French Fifth Republic." *American Political Science Review* 67:29–54.

————. 1977. "Spatial Voting Models for the French Fifth Republic." *American Political Science Review* 71:1447–66.

Roth, G. 1963. *The Social Democrats in Imperial Germany*. Totowa, N.J.: Bedminster Press.

Rubinstein, A. 1986. "Finite Automata Play the Repeated Prisoners' Dilemma." *Journal of Economic Theory* 39:83–96.

Rudd, C. 1986. "Coalition Formation and Maintenance in Belgium." In G. Pridham (ed.), *Coalition Behaviour in Theory and in Practice: An Inductive Model for Western Europe*. Cambridge: Cambridge University Press.

Rudner, R. 1966. "On the Objectivity of Social Science." *Philosophy of Social Science*. Englewood Cliffs, N.J.: Prentice-Hall.

Runge, C. F. 1984. "Institutions and the Free Rider: The Assurance Problem in Collective Action." *Journal of Politics* 46:154–81.

Rush, M. 1969. *The Selection of Parliamentary Candidates*. London: Nelson and Sons.

Sabel, C. F. 1981. "The Internal Politics of Trade Unions." In S. Berger (ed.), *Organizing Interests in Western Europe*. Cambridge: Cambridge University Press.

Samuelson, P. A. 1954. "The Pure Theory of Public Expenditure." *Review of Economics and Statistics* 36:378–89.

———. 1963. "Problems of Methodology—Discussion." *American Economic Review Papers and Proceedings* 53:231–36.

Sartori, G. 1976. *Parties and Party Systems*. Cambridge: Cambridge University Press.

———. 1978. "Anti-elitism Revisited." *Government and Opposition* 13:58–80.

Savage, L. J. 1954. *The Foundations of Statistics*. New York: Wiley.

Scharpf, F. W. 1987. "A Game-Theoretical Interpretation of Inflation and Unemployment in Western Europe." *Journal of Public Policy* 7:227–57.

Scharpf, F. W., and Ryll, A. 1988. "Core Games, Connected Games, and Networks of Interaction." Cologne: Max-Planck-Institut für Gesellschaftsforschung. Mimeo.

Schattschneider, E. E. 1960. *The Semi-sovereign People*. New York: Holt, Reinhart and Winston.

Schlesinger, J. A. 1984. "On the Theory of Party Organization." *Journal of Politics* 46:369–400.

Schmitt, D. E. 1974. *Violence in Northern Ireland: Ethnic Conflict and Radicalization in an International Setting*. Morristown, N.J.: General Learning Press.

Schotter, A. 1981. *The Economic Theory of Social Institutions*. Cambridge: Cambridge University Press.

Schumpeter, J. A. 1947. *Capitalism, Socialism, and Democracy*. New York: Harper and Row.

Schwartz, T. 1982. "No Minimally Reasonable Collective Choice Process Can Be Strategy-Proof." *Mathematical Social Science* 3:57–72.

———. 1985. "The Meaning of Instability." Paper presented to the American Political Science Association meetings.

Schwartzenberg, R. G. 1979. *Politique Comparée*. Paris: Les Cours de Droit.

Sklar, R. L. 1963. *Nigerian Political Parties: Power in an Emergent African Nation*. Princeton, N.J.: Princeton University Press.

———. 1979. "The Nature of Class Domination in Africa." *Journal of Modern African Studies* 17:531–32.

Scott, J. C. 1976. *The Moral Economy of the Peasant*. New Haven, Conn.: Yale University Press.

Scriven, M. 1962. "Explanations, Predictions, and Laws." In H. Feigl and G. Maxwell (eds.), *Minnesota Studies in the Philosophy of Science*, vol. 3. Minneapolis, Minn.: University of Minnesota Press.

Selten, R. 1975. "Reexamination of the Perfectness Concept for Equilibrium Points in Extensive Games." *International Journal of Game Theory* 4:25–55.

———. 1978. "The Chain Store Paradox." *Theory and Decision* 9:127–59.

Sen, A. K. 1967. "Isolation, Assurance and the Social Rate of Discount." *Quarterly Journal of Economics* 81:112–24.

Shannon, W. W. 1968. *Party, Constituency and Congressional Voting.* Baton Rouge: Louisiana State University Press.

Shapley, L. S., and Shubik, M. 1954. "A Method for Evaluating the Distribution of Power in a Committee System." *American Political Science Review* 48:787–92.

Shepsle, K. A. 1979. "Institutional Arrangements and Equilibrium in Multidimensional Voting Models." *American Journal of Political Science* 23:27–59.

———. 1986. "Institutional Equilibrium and Equilibrium Institutions." In H. Weisenberg (ed.), *Political Science: The Science of Politics.* New York: Agathon Press.

Shepsle, K. A., and Cohen, R. N. 1988. "Multiparty Competition, Entry, and Entry Deterrence in Spatial Models of Elections." In J. Enelow and M. Hinich (eds.), *Readings in the Spatial Theory of Voting.* Cambridge: Cambridge University Press.

Shepsle, K. A., and Weingast, B. R. 1981. "Political Preferences for the Pork Barrel: A Generalization." *American Journal of Political Science* 25:96–111.

———. 1984. "Uncovered Sets and Sophisticated Voting Outcomes with Implications for Agenda Institutions." *American Journal of Political Science* 28:49–74.

———. 1987. "Institutional Foundations of Committee Power." *American Political Science Review* 81:85–104.

Shubik, M. 1982. *Game Theory in the Social Sciences.* Cambridge, Mass.: MIT Press.

Simon, H. A. 1957. *Models of Man.* New York: Wiley.

Skyrms, B. 1986. *Choice and Chance: An Introduction to Inductive Logic.* Belmont, Calif.: Wadsworth.

Snyder, G. H. 1971. "Prisoners' Dilemma and Chicken Games in International Politics." *International Studies Quarterly* 15:66–103.

Snyder, G. H., and Diesing, P. 1977. *Conflict Among Nations: Bargaining, Decision-Making and System Structure in International Crises.* Princeton, N.J.: Princeton University Press.

Steiner, J. 1969. "Conflict Resolution and Democratic Stability in Subculturally Segmented Political Systems." *Res Publica* 11:775–98.

———. 1974. *Amicable Agreement Versus Majority Rule: Conflict Resolution in Switzerland.* Chapel Hill, N.C.: University of North Carolina Press.

Stinchcombe, A. L. 1980. "Is the Prisoners' Dilemma All of Sociology?" *Inquiry* 23:187–92.

Stouffer, S. A. 1965. *The American Soldier*. New York: Wiley.

Strom, K. 1984. "Minority Governments in Parliamentary Democracies: The Rationality of Nonwinning Cabinet Solutions." *Comparative Political Studies* 17:199–228.

Taverne, D. 1974. *The Future of the Left*. London: Jonathan Cape.

Taylor, C. 1965. "Interpretation and the Sciences of Man." *Review of Metaphysics* 25:3–51.

Taylor, M. 1976. *Anarchy and Cooperation*. New York: Wiley.

Taylor, M., and Ward, H. 1982. "Chickens, Whales, and Lumpy Goods: Alternative Models of Public-Goods Provision." *Political Studies* 30:350–70.

Theil, H. 1968. *Optimal Decision Rules for Government and Industry*. Amsterdam: North-Holland.

Thompson, E. A., and Faith, R. L. 1981. "A Pure Theory of Strategic Behavior and Social Institutions." *American Economic Review* 71:366–80.

Thung, M. A., Peelen, G. J., and Kingmans, M. C. 1982. "Dutch Pillarization on the Move? Political Destabilization and Religious Change." *West European Politics* 5:127–48.

Trockel, W. 1986. "The Chain-Store Paradox Revisited." *Theory and Decision* 21:163–79.

Truman, D. 1951. *The Governmental Process*. New York: Knopf.

Tsebelis, G. 1985. "Parties and Activists." Ph.D. thesis, Washington University.

———. 1986. "A General Model of Tactical and Inverse Tactical Voting." *British Journal of Political Science* 16:395–404.

———. 1988a. "Nested Games: The Cohesion of French Electoral Coalitions." *British Journal of Political Science* 18:145–70.

———. 1988b. "When Do Allies Become Rivals?" *Comparative Politics* 20:233–40.

———. 1989. "The Abuse of Probability in Political Analysis: The Robinson Crusoe Fallacy." *American Political Science Review* 83:77–91.

———. Forthcoming. "Are Sanctions Effective? A Game Theoretic Analysis." *Journal of Conflict Resolution*.

Tucker, A. 1950. "A Two-Person Dilemma." Mimeo, Stanford University.

Turner, J. 1951. *Party and Constituency: Pressures on Congress*. Baltimore: Johns Hopkins University Press.

Tversky, A., and Kahneman, D. 1981. "The Framing of Decisions and the Psychology of Choice." *Science* 211:453–58.

Van Damme, E. E. C. 1984. *Refinements of the Nash Equilibrium Concept.* New York: Springer-Verlag.

Verba, S. 1965. "Comparative Political Culture." In L. Pye and S. Verba (eds.), *Political Culture and Political Development.* Princeton, N.J.: Princeton University Press.

Welch, S., and Studlar, D. T. 1983. "The Policy Opinions of British Political Activists." *Political Studies* 31:604–19.

Whiteley, P. 1983. *The Labour Party in Crisis.* London: Methuen.

Williams, P. 1983. "The Labour Party: The Rise of the Left." *West European Politics* 6:27–55.

Williamson, O. E. 1985. *The Economic Institutions of Capitalism.* New York: Free Press.

Wright, V. 1983. *The Government and Politics of France.* New York: Holmes and Meier.

Young, A. 1983. *The Reselection of MPs.* London: Heinemann Educational Books.

Index